The Rise of Empires

Sangaralingam Ramesh

The Rise of Empires

The Political Economy of Innovation

palgrave
macmillan

Sangaralingam Ramesh
Department for Continuing Education
University of Oxford, Rewley House
Oxford, UK

ISBN 978-3-030-01607-4 ISBN 978-3-030-01608-1 (eBook)
https://doi.org/10.1007/978-3-030-01608-1

Library of Congress Control Number: 2018957627

Cover illustration: GettyImages/whitemay

This Palgrave Macmillan imprint is published by the registered company Springer Nature Switzerland AG
The registered company address is: Gewerbestrasse 11, 6330 Cham, Switzerland

For my father and my mother,
Nallathamby Sangaralingam and Pathmarani Sangaralingam,
Inuvil and Karinagar, Ceylon.
For their love, compassion and their humanity

Preface

As I wrote 'China's Lessons for India', Volume 1 and Volume 2, published by Palgrave Macmillan in September/October 2017, my interest in the process of innovation and its outcomes was stimulated. This is specifically with regards to how innovation may give one society a competitive advantage over another, leading to the 'Rise of Empires'. Contextually, my interest was also stimulated by the economic and political rise of China and the impending rise of India. The rise of China, with its Belt and Road Initiative, investments in the infrastructure of other countries, the increase in its capacity of technological innovation, resembles the rise of an empire. In countries such as Sri Lanka, an ex-British colony, local people are learning Mandarin to work with and for Chinese companies working on Chinese-funded infrastructure projects on the island. India has become encircled by China's economic embrace of its neighbours. An independent India and China have fought wars related to border disputes. Recently, trouble flared on the Doklam Plateau, an area of territorial dispute between Bhutan and China. In the last five years, China has also built artificial islands in the South China Sea. In the light of history and its tendency to repeat itself, some might ask if China is heading in the same direction, in the coming decades, as Japan following the Meiji Restoration in 1868. Following its economic rise, Japan needed resources which it did not have. The only way it could continue to sustain its economic growth was to acquire resources from other countries, much more cheaply, by using military force and occupation. This it did by invading China and occupying Manchuria. However, once the United States imposed trade embargoes on Japan, its economy was strangled. This precipitated Japan's attack on the the United States naval base at Pearl Harbour, with which the United States entered World War II. However, in ancient times, the kingdoms of the various parts of the Indian subcontinent had close filial connections with China and its dynasties. The Chola Empire of eleventh-century southern India engaged in maritime trade with the Tang, Song, and the Yuan dynasties of China. At that time, Europe was still undeveloped and all the latterly colonised lands, Canada, America, South America, Australia, New Zealand, India, Indonesia, Africa, and other parts of the world, were still controlled and habituated by their native peoples. The centre of the world was in Asia. Today, the world is again at a turning point, with the axes of economic and political power once again shifting from the west to the east. Britain and the United States were long seen to be the dominant powers in the world. Britain retained its superpower status until the early twentieth century, when it was overtaken by an ascendant United States, its former colony. Now, the democratic systems of both countries have led them to a state of decline, in comparison to China and India. Moreover, in comparison to fifty years ago, the United States and Great Britain do not have the same calibre of politicians. This is especially true in the context of integrity, intellect, and sincerity. For example, Tony Blair quipped that he thought there would be enough economic growth to absorb workers from Eastern Europe who arrived because of his government's acquiescence of the EU's free movement of labour. Nevertheless, at a theoretical economic level, the free movement of labour would only have been required for countries with a common currency. In other words, those countries which had adopted the Euro in 1999.

The decline of Britain and the United States may have happened for many reasons, but economically, this decline can be attributed to a wasteful use of resources. Politicians strive and contrive to win power for their own good rather than the good of the people. They make promises and follow economic policies that have no economic sense behind them. One example is Britain's decision to join the EEC in 1973. The voters thought that it was a trading relationship. However, the politicians knew otherwise.

They knew that eventually, all the countries in the subsequent EU would form the United States of Europe. In the United States, the resources wasted and the lives lost on both sides during the Vietnam War serve as an example of our lack of humanity as well as a lack of intellect. In the light of history, if the United States had left Vietnam to itself, like all the other planned economies did, Vietnam would have turned by itself to embrace Capitalism. The United States wasted a valuable opportunity in not pursuing the exploration and the colonisation of space. With appropriate levels of funding, this objective would have driven the United States to become a more highly innovative economy than it was. And the country's economic growth would have increased substantially. It would also have furthered the dreams of humanity and its long-term survivability.

The purpose of this book is to identify the process by which innovation takes place, the factors responsible for this, how empires subsequently rise, and how they fall. This will be accomplished by analysing the rise of empires over the last 8000 years at a multicultural level, starting with the emergence of Homo Sapiens. Furthermore, the book takes an interdisciplinary approach, and so its findings will be relevant to economics, anthropology, political economy, politics, and international relations.

Oxford, UK Sangaralingam Ramesh

[illegible]

Acknowledgements

I would like to thank Rachel Sangster, the Global Head of Economics and Finance, as well as Joseph Johnson at Palgrave Macmillan for all their invaluable help in getting this book published. I would also like to thank the library service at University College London, where the idea for this book was formulated in October 2017, and where the writing of the book began in January 2018.

Contents

1
Introduction

The purpose of this book is to evaluate the role of innovation in economics, the factors it depends on, and how it emerges over time, based on a historical analysis of several civilisations over a 10,000-year period starting with the emergence of humanity from its place of development and evolution in Africa. In this case, the analysis and evaluation begin with the emergence of Homo Sapiens from Africa, and extends to the Palaeolithic and to the Bronze Age. The earliest contemporary civilisations, ancient Mesopotamia, ancient Egypt, and the Indus Valley civilisation of modern-day northern India are also discussed. While the civilisations of ancient Mesopotamia and ancient Egypt proved successful and relatively longer lasting than the Indus Valley civilisation, the latter was vast and, perhaps, more technologically, economically, and politically advanced than the others. Contextually, the Indus Valley civilisation covered an area comparable to a quarter of the size of Western Europe, populated by an estimated one million people, equivalent to the size of the population of Rome at the height of its power.[1] Furthermore, archaeological evidence suggests that the Indus Valley civilisation

[1] Robinson, A. (2015), The Indus: Lost Civilisations, Reaktion Books Ltd., London.

S. Ramesh, *The Rise of Empires*, https://doi.org/10.1007/978-3-030-01608-1_1

constituted 80,000 settlements ranging from villages to towns to cities, of which there were five major sites, two of the biggest of these being Harappa and Mohenjo Daro.[2] All three civilisations traded with each other, contributing to a diffusion of knowledge between its peoples and others. However, the more advanced Indus Valley civilisation died out while the other two civilisations lasted. The Indus Valley civilisation was considered advanced for its time for a number of reasons.[3] Firstly, the civilisation built merchant ships that enabled it to trade with the rest of the ancient world, including the other major contemporary civilisations of the time—ancient Mesopotamia and ancient Egypt. Secondly, the civilisation had the technological capacity to build drainage and sanitation systems comparable to ancient Rome, but 2000 years ahead of its time.[4] Thirdly, the major metropolises of the civilisation featured well-planned streets, buildings made of mud baked bricks—a first for the ancient world—and major storage facilities for water and grain. Lastly, and perhaps more indicative that the Indus Valley civilisation represented a highly developed ancient economy, was the use of standardised weights that utilised a binary/decimal system.[5] The use of such a system would allow for the weighing, distribution, and storage of agricultural produce. It would also facilitate a system of payments to soldiers and workers by the relevant authorities. However, the Indus Valley civilisation died out. A number of reasons have been put forward for the early demise of the Indus Valley civilisation, ranging from attacks by hostile tribes attracted by its prosperity to natural disasters such as earthquakes and the changing flows of rivers.[6] The latter is most likely because most of the urban centres of the Indus Valley civilisation were located next to rivers. Nevertheless, the longevity of the civilisations of ancient Egypt and ancient Mesopotamia may have indirectly contributed towards the development of civilisation in the Mediterranean. The Babylonian Empire

---

[2] Ibid.

[3] Ibid.

[4] Ibid.

[5] Ibid.

[6] Chaurasia, R. (2002), History of Ancient India, Earliest Times to 1200 AD, Atlantic Publishers & Distributors (P), Ltd, New Delhi.

formed upon the remnants of ancient Mesopotamia. It too traded with the peoples of the Mediterranean. It was later in this region that the civilisation of ancient Greece developed, with its city states and the development of the contemporary western ideal of freedom of the individual and property rights. Thus, the basic tenets of contemporary traditional neoclassical economics, freedom, trade, and private property rights lie in this ancient world. The demise of the civilisation of ancient Greece was superseded by the rise of the Roman Republic, which itself was superseded by the Roman Empire. The ideals of governance of the Roman Republic, the freedom of the citizen, and governance by the citizen would not be seen in the world again until the founding of the American Republic in the eighteenth century AD. However, the Roman Republic met its demise with the rise of populism and a lurch towards autocracy, and the rise of the Roman Empire. This has alarming parallels with the state of the contemporary world. The analysis then moves to the Chola Dynasty, which existed in the southern part of contemporary India. The empire of the Chola Dynasty included modern-day Indonesia, Sri Lanka, Malaysia, and Cambodia. The Chola Empire traded with the Roman Empire, which had trading posts in southern India as well as with the Song Dynasty of China. The economy of the latter was built on trade. However, it was during the period of the Han Dynasty that paper and printing were invented, with gunpowder being invented in the ninth century AD. This was just before the period that encompassed the Song Dynasty, which may have been the first civilisation in the world to issue paper money. Furthermore, it was during the Song Dynasty that the world first witnessed an economic crisis due to inflation, as the Song government printed money to finance its defences. In this case, the Chinese were the first to recognise the relationship between the money supply and the price level.[7] The Song Dynasty came close to leading China to be the first country in the world to experience industrialisation due to its economic prosperity through trade. However, Imperial China

[7] Goetzmann, W., and Koll, E. (2005), Paying in Paper, A Government Voucher from the Southern Song, IN The Origins of Value: The Financial Innovations That Created Modern Capital Markets, Oxford University Press, Oxford.

did not share the two cornerstones of the European market economy, which would have allowed it to foster entrepreneurship through incentivisation. Nevertheless, prices in ancient China were not set by the government, but by the market forces of supply and demand. The first of these cornerstones was the right of the individual to own property as well as to make legally enforceable contracts.[8] The second cornerstone, which was absent in ancient China, was the lack of commercial institutions formalised and protected by the state's legal institutions.[9] In this context, the institutions which formed from the end of the Han Dynasty to the beginning of the Song Dynasty, while supporting a complex market as the one which existed in Medieval Europe, were of a completely different nature compared to their western counterparts. In this regard, the main institution that facilitated the maintenance of economic harmony in ancient China, was the hereditary structure of society.[10] This particularly reflected in a social organisation characterised as being composed of the Imperial household and the nobility. The nobility owned all of the land, had access to education, and accumulated all of the wealth of Chinese society.[11] Furthermore, the nobility intermarried. The purpose of this setup was to ensure a smooth succession from one generation of the nobility to the next, ensuring that commoners remained commoners and were prevented from moving out of the category of their birth.[12] The purpose of government was, therefore, to regulate the economic activities of people so that this status quo could be maintained.[13] In order to ensure this, certain areas of cities were restricted solely for the occupation of merchants and artisans and, in these areas, court officials would take charge of all aspects of manufacturing and business.[14] However, from the

[8] Hamilton, G. (1996), The Organisational Foundations of Western and Chinese Commerce: A Historical and Comparative Analysis, IN Asian Business Networks, Hamilton, G. (Ed), Walter de Gruyter, Berlin.

[9] Ibid.

[10] Ibid.

[11] Ibid.

[12] Ibid.

[13] Ibid.

[14] Hamilton, G. (1996), The Organisational Foundations of Western and Chinese Commerce: A Historical and Comparative Analysis, IN Asian Business Networks, Hamilton, G. (Ed), Walter de Gruyter, Berlin.

eighth century AD to the thirteenth century AD, through the period of the Song Dynasty, the power of the nobility began to decline, and the social mobility of the commoner began to become more fluid.[15] During the Song Dynasty, control of manufacture and business by merchants and artisans became indirect, with the formation of guilds through which some form of court supervision and regulation was practised.[16] These guilds, while replacing the market economy which characterised the Tang Dynasty, were more suited to an urban economy.[17] The Imperial government benefited more from this arrangement by being able to collect more tax revenues due to the increased economic prosperity. While the Song Dynasty was ended by the Mongol invasion and the usurpation of imperial power by the Yuan Dynasty, the economic development of institutions, such as that of the merchant and artisan guilds, ended in China. However, merchant guilds also formed in England from the late thirteenth century,[18] and they continued to develop over the next few centuries because, while the Mongols reached Hungary, Poland, and Germany, they turned back to Mongolia before reaching the shores of England. Thus, the development of the guilds would ensure that England, through its Mercantile policies, would set the stage for the formation of its empire, starting in the sixteenth century. However, this would reach a peak with England being the first country to experience the Industrial Revolution, starting in the late eighteenth century and continuing through the nineteenth century. The remnants of the British Empire would last until the latter part of the twentieth century, ending with the handover of Hong Kong back to China in 1997. Interceding the years between the demise of the Song Dynasty and the rise of the British Empire was the Mongol Empire. The latter would diffuse knowledge, invention, and innovation from east to west, in effect, stimulating the rise of the British Empire. This knowledge would encompass gunpowder, the compass, paper, and

[15] Ibid.

[16] Ibid.

[17] Ibid.

[18] Rosser, G. (2015), The Art of Solidarity in the Middle Ages: Guilds in England 1250–1550, Oxford University Press, Oxford.

printing. The British Empire emerged and remained strong and large, in contrast to the empires of other European powers, because it embraced technological innovation and political innovation. Furthermore, the rise of the British Empire was based on state capitalism in the context of trading monopolies with specific regions of the world granted by the English Crown to various regionally specialised English companies. This was an English financial and corporate governance innovation which incentivised entrepreneurship and made the accumulation of substantial financial resources by a trading company much easier to achieve. In conjunction with the doctrine of Mercantilism, this gave rise to the British Empire. The rise of which has stark similarities to the rise of China today. China's trade policies are akin to English Mercantilism, the acquisition of assets in other countries by financial largesse. Loaning money to countries by, for example, building infrastructure, allowing these countries to become indebted to China, and only settling loans in return for physical assets—such as ports. This was what happened in Sri Lanka and Pakistan.

Traditionally in economics, innovation is regarded as facilitating long-term economic growth based on two schools of thought that included the Solow Model, in which technology is assumed to be freely available to firms, and the endogenous model of economic growth in which innovation and technological change is assumed to arise from within firms. While the former model does not address how innovation occurs or the factors responsible for it, the latter places an emphasis on education, learning by doing, and the diffusion of knowledge as contributing to innovation. However, neither model specifically addresses the historical development of innovation, the sources of innovation, its impact on the development of economic systems, as well as the historical impact of innovation on human economic activity. This book seeks to fill this gap by evaluating how innovation has developed at a historical level as well as at a species level based on the idea that Homo Sapiens maximise survival and not utility. Nevertheless, the fundamental issue that economics attempts to resolve is how unlimited wants can be met by scarce resources, and not the maximisation of survival. This paradox gives rise to choice. Consumers must choose what to consume and what not to consume, and firms must decide upon what to produce and what not to produce. Theoretical economics, in order to provide a rigorous basis for analysis, asserts that consumers maximise utility or satisfaction from consumption, whereas a firm's production decisions are based on the maximisation of profit. Moreover, according to standard economic theory, consumers are deemed to make 'rational' choices/decisions. In other words, choices or decisions are made on the basis of pure logic, and in no way depend on our emotions.

The concept of utility championed by neoclassical economics has its roots in the works of Aristotle and ancient Greek economic theory. Aristotle was the starting point in western economic thinking of an attempt to determine a theory of value. Aristotle thought of value in terms of use. In other words, people would seek to possess 'things' that give them the greatest level of satisfaction and/or productivity. An individual in a hunter-gatherer society may assert great value with a knife, spear, bow and arrow, fire, and some form of cooking implements. Without these things, it would be difficult to acquire meat, to cook it, and to eat it. However, an individual in a hunter-gatherer society may give less value to plant crops that could be picked with more ease and that could be more abundant. Exchange value, on the other hand, arises when meat and plant crops may be traded either within the members of a hunter-gatherer tribe or between such tribes. In a non-monetised economy, such trade would take place through barter. But what would be the basis for the determination of value through a process of barter? Herein, Adam Smith saw a problem arising from the distinction between 'use value' and 'exchange value', and he called it a Paradox of Value. On the basis of its subjectivity, Smith rejected the notion of 'use value', and instead placed greater emphasis on 'exchange value'. In Smith's eyes, the latter could be determined based on the cost of production. For example, the exchange value of a beaver pelt and deer meat could be established based on how many hunters were required to 'produce' each. Neoclassical economics, on the other hand, extends Aristotle's 'use value' to total utility, marginal utility, and the derivation of the demand curve. In all its theoretical deliberations, economics tends to be missing the elements of nature, that is, as human beings we are an animal—a mammal, to be more specific. However, Alfred Marshall did recognise a hierarchy of human needs, each distinguishable from the other.[19] In this case, Marshall recognised the biological needs associated with food, clothing, and shelter.[20] However, he failed to recognise that the economic objective of human activity is to maximise survival, and this is transformed to different levels by invention and innovation over time. Moreover, to all extents and purposes, these human needs have not been explicitly recognised by theoretical economics. Like all animals, we tend not to maximise utility or satisfaction, but we do want to maximise our chances of survival to pass our genes onto subsequent generations. In essence, this is what drives our economic behaviour. Furthermore, economic behaviour is dependent on human cognition. In conjunction with our cognitive intelligence, the maximisation of survival facilitates invention, and then this leads to innovation. An invention is a new idea that can be tangible (technological) or intangible (an economic, political, religious, or cultural doctrine). On the other hand, an innovation builds upon an invention to produce something new through further invention. The artificial creation of fire by one of Homo Sapiens' earlier ancestors, Homo Erectus was an invention based on the observation of a naturally observing phenomena. But the use of fire by Homo Erectus to cook meat was an innovation. This unknowingly contributed towards evolution in increasing brain size and cognitive powers, in the transformation of Homo Erectus into Homo Sapiens. However, while Homo Erectus had spread across the continents before the emergence of Homo Sapiens, it was only in Africa that Homo Erectus evolved into Homo Sapiens.[21] In other parts of the world in which Homo Erectus existed, the species became an evolutionary dead end and died out. Another example of invention and innovation would be the motor car. For instance, the motor car is an innovation that built on the invention of the wheel through the invention of the steam engine. Further innovation took place when the steam engine was replaced by an engine fuelled by petrol, a substance derived from oil. The process of refining oil to

---

[19] Haines, W. (1982), Psychoeconomics of Human Needs: Maslow's Hierarchy and Marshall's Organic Growth, Journal of Behavioral Economics, Vol. 11 (Winter), pp. 97–121.

[20] Ibid.

[21] Nanjira, D. (2010), African Foreign Policy and Diplomacy from Antiquity to the 21st Century, Praeger, Santa Barbara, California.

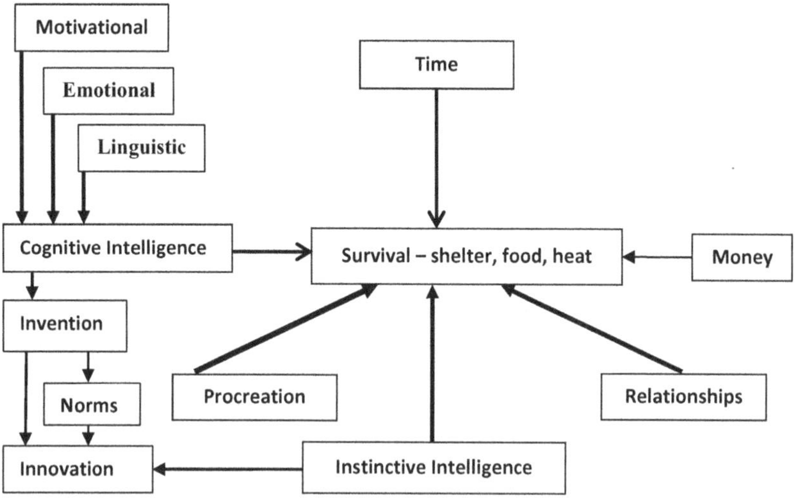

Fig. 1.1 Homo Sapiens and animals—Factor analysis. (Source: Author)

produce petrol and other derivatives was itself an invention. The idea that petrol could fuel an engine, resulting in increased efficiency, but with more damage to the environment, was in itself an innovation. Cultural norms may also facilitate invention and innovation. For example, the emergence of the environmental sustainability trend in the latter half of the twentieth century led to new ways of thinking regarding power sources (solar panels), production processes, and packaging (use of plastics), as well as the natural world (deforestation, species extinction and ecosystem destruction).

To put this into context, since the emergence of civilisation in ancient Sumer, whilst Homo Sapiens has represented only 0.01% of all life on Earth, it has destroyed 83% of all wild mammal species as well as 50% of all plant species.[22]

Figure 1.1 above represents the schematic framework which shows the common and the uncommon factors that allow for the survival of Homo

[22] Bar-On, Y., Phillips, R., and Milo, R. (2018), The biomass distribution on Earth, Proceedings of the National Academy of Sciences of the United States of America, www.pnas.org/cgi/doi/10.1073/pnas.1711842115

Sapiens as well as other animals. While the nature of the universe is such that it is governed by the fundamental laws of physics and mathematics, human beings are the results of the universe and more irrational than the apparent rationality of the universe. Moreover, man may be an enlightened animal, but he still shares with animals the need to survive. In the short term, on an individual basis, this can be achieved by seeking food and shelter. In the long run, however, for the species to survive, the individual must not only survive, but also procreate. Procreation also facilitates evolution through the biological pathways of genetic change. As an individual adapts to his/her environment biological metabolic pathways facilitate the production of proteins by the body, which results in subtle changes to the individual's RNA and DNA over a lifetime, although the changes which take place when the individuals are reproductively active are the most important, as these changes are passed to offspring, and this elicits the evolution of the species over time. As the rate of technological change impacts on an individual environment, it also impacts on the evolutionary change of Homo Sapiens.

The factors which are needed for survival—such as shelter, food, and procreation—occur over time. This is to say that the requirement for the factors of survival must be met daily and on a longer-term basis respectively. This would of course depend on the needs of the species. However, animals can be innovative through their 'instinctive intelligence'. For example, recently, there was a case of a crayfish detaching its claw to escape from being part of a boiling soup.[23] Homo sapiens are distinct from other species in two respects. Firstly, Homo sapiens' cognitive abilities set it apart from other species, giving it unique behavioural advantages linked to the use of objects as well as the use of language and writing. These cognitive abilities may give rise to our greater capacity to invent and innovate. Secondly, Homo sapiens are the only species that uses money. The allure of money is such that its presence makes Homo Sapiens act irrationally due to greed rather than logic. While Homo sapiens'

[23] Guardian (2018), Crayfish becomes online hero by detaching claw to escape boiling soup, https://www.theguardian.com/world/2018/jun/03/crayfish-becomes-online-hero-by-detaching-claw-to-escape-boiling- soup

cognitive ability will allow it to survive in a monetised and a non-monetised economy, the availability of and access to money will help man survive in a monetised economy. The distinction between a non-monetised economy and a monetised economy extends to the distinction between a hunter-gatherer existence and an urban, centred existence.

The shift of humanity from a hunter-gatherer existence to an agricultural one also represented a shift from a rural to an urban existence in which people were concentrated and their numbers rising. A constant and more reliable source of food, represented by agriculture and animal husbandry, allowed substance and sustenance for more 'leisure' time and time to engage the cognitive capacity of the human mind. This allowed for more thinking time, which led to increasing innovation. Innovations represent responses to the internal and external environments faced by an entity.[24] However, an innovation could also be adopted as a pre-emptive action to influence the environment.[25] In this case, innovations allow entities (organisations, society, state, and the individual) to evolve with their environments to achieve a 'best fit'.[26] There is also a positive relationship between the size of an entity and the level of innovativeness.[27] This could simply be because more resources are available, and the use of those resources are better organised. Innovation leads to a society, state, individual, or organisation having a comparative advantage over another.[28] However, the accumulation of knowledge by states and civilisations over time allows for greater innovation. For example, the invention of gunpowder by unknown Chinese alchemists in the ninth century AD, and the diffusion of the knowledge of its composition to the west due to the Mongolian invasions of Europe of the thirteenth century AD greatly changed the face of warfare. The development of the cannon and the

[24] Damanpour, F. (1992), Organisational Size and Innovation, Organisation Studies, 13/3, pp. 375–402.

[25] Ibid.

[26] Ibid.

[27] Ibid.

[28] Helfat, C. E., Finkelstein, S., Mitchell, W., Peteraf, M., Singh, H., Teece, D. and Winter, S. (2007), Dynamic Capabilities: Understanding Strategic Change in Organizations. New York: Wiley.

musket changed the rules of engagement, on both land and sea. Ships with more and efficient cannons could defeat ships with inferior or no cannons at a distance. Cannon and musket would allow an army to denigrate another, allowing for the final fatal engagement of sword-wielding cavalry. The states of Europe would jockey for world supremacy using cannon and musket from the sixteenth century onwards, until the mid-twentieth century AD.

Innovation can be thought of both as a process and as an outcome. Innovation can be technological or conceptual. Technological innovation can lead to better ways of doing things. For example, the development of writing in ancient Sumer allowed for the storage of information on clay tablets. Knowledge could then be transmitted from one generation to the next, allowing knowledge to accumulate, scholars to learn, ideas to form, and innovations to occur. Ancient Sumer is also reputed to be the civilisation in which the wheel was invented, as well as the civilisation that conceptualised the division of the day into hours and minutes. The invention of the wheel could have occurred by observing the movement of a stone rolling down a hill. Just as centuries later, man would observe the flight of birds and develop aircraft in the early part of the twentieth century AD, in ancient Sumer, the rolling effect of pebbles, rocks, and boulders down a hill may have given an ancient Sumerian the inquisitive initiative to mimic the same effect with other materials, such as wood, to substitute the effect for a useful purpose. The early wheel may have started with planks nailed together to imitate an imperfect circle. Over time, with developments in wood working and metallurgy, a wheel which rotated perfectly around 360 degrees would have been developed. The wheel would have allowed for the transportation of more goods, building materials, and large numbers of people over longer distances. Centuries later, in the late nineteenth and twentieth centuries AD, the wheel would be conjoined to a motor fuelled by fossil fuels to form a motor car. This example shows how the accumulation and diffusion of knowledge allows for the innovation upon invention centuries apart. The invention of the wheel would also have allowed trade between cities to increase, facilitating and expanding markets. This would integrate more people and allow for the diffusion of knowledge between different civilisations. Furthermore, more buildings could be built faster and soldiers transported to trouble spots more quickly, leading to greater state control and stability. The division of the day into hours and minutes would have allowed for the better management of time, as well as leading to a more efficient organisation of activities. A favourable cultural environment facilitates innovation.[29] Moreover, the positive effects of laws and institutions on governance, promoting economic and civil stability, has been found to have a positive correlation on economic growth.[30] However, economic growth results from innovation.[31] Therefore, it must logically follow that laws and institutions have a direct correlation with the level of innovation. Although this may be the case, there is little direct theoretical or empirical proof of how the level of governance affects the relationship between laws, institutions, and innovation.[32] Nevertheless, there is some empirical support for the idea that national systems of innovation are driven by the simultaneous development of innovative variables such as creativity, research, and the innovation and the institutional variables of infrastructure, trade, and human capital.[33]

The outcome of innovation can be something tangible, like a war chariot and writing, or it can be intangible like a political, socioeconomic practice, or a custom. Innovations can influence both human cognitive ability as well as on human productivity. For example, the innovation of writing allowed for thoughts to be better organised and recorded so that they could be developed further later through more thinking. This would allow options to be considered and strategies to be developed. Learning would become more effective, increasing the possibilities for innovation.

---

[29] Crossan, M., and Apaydin, M. (2010), A Multi-Dimensional Framework of Organisational Innovation: A Systematic Review of the Literature, Journal of Management Studies, 47:6.

[30] La Porta, R., Lopez-de-Silanes, F., Shleifer., A and Vishny, R. (2000), Investor Protection and Corporate Governance, Journal of Financial Economics, 58, 3–27.

[31] Aghion, P., and P. Howitt, P. (2006), Appropriate Growth Policy: A Unifying Framework, Journal of the European Economic Association, 4 (2006), 269–314.

[32] Sapra, H., Subramanian, A., and Subramanian, K. (2014), Corporate Governance and Innovation: Theory and Evidence, Journal of Financial and Quantitative Analysis, Vol. 49, No. 4, pp. 957–1003.

[33] Castellacci, F., and Natera, J. (2013), The dynamics of national innovation systems: A panel co-integration analysis of the coevolution between innovative capability and absorptive capacity, Research Policy, 42, pp. 579–594.

The state could also maintain records of its citizens, making tax collection more effective. On the other hand, the war chariot would allow soldiers to travel faster over long distances, taking more weapons and supplies than they could on foot. The soldiers would also have a higher vantage point from which to throw spears and launch arrows at an enemy based on foot. Innovations in political, socioeconomic practice and customs could bring about changes in the organisation of society. For example, the caste system in India evolved over thousands of years following the demise of the Indus Valley civilisation. The caste system groups people into societal strata that they join upon birth and do not leave until they die. The main castes of the Indian caste system, in hierarchical order, are the Brahmins, the Kshatriyas, the Vaisyas, and the Sudras. The Brahmins are the priestly, scholarly learning caste at the top of the caste hierarchy.[34] The Brahmins are followed by the Kshatriyas or warrior class, whose function is to protect society.[35] The Kshatriyas are followed by the Vaisyas, who occupy the productive part of the economy, undertaking activities associated with the rearing of cattle, money lending, trade, and crop cultivation.[36] The final member of the caste system is the Sudras caste, whose only function was the servicing of the upper castes.[37] While India's caste system organised its society, it was an intangible innovation with tangible results, the impact on people's lives by having been born into the Sudras class, for example. Thousands of years after the roots of India's caste system were laid, the political innovation of Mercantilism emerged in Europe, England, and Prussia in the eighteenth century. In the case of England, Mercantilism laid the roots of the British Empire whereas, in the case of Prussia, Mercantilism led to the unification of the independent German states into a greater Germany. With regards to Mercantilism, it may be said that there is a causal relationship between context, policy,

[34] Kumari, A., and Bhaskar, S. (1998), Social Change Among Balijas, Majority Community of Andhra Pradesh, MD Publication PVT Ltd., New Delhi.

[35] Ibid.

[36] Ibid.

[37] Ibid.

and innovation sequentially.[38] As markets grew, merchants became prosperous, and more financial resources became available to finance invention and innovation.

References

Aghion, P., and P. Howitt, (2006), Appropriate Growth Policy: A Unifying Framework, Journal of the European Economic Association, 4 (2006), 269–314.

Bar-On, Y., Phillips, R., and Milo, R. (2018), The biomass distribution on Earth, Proceedings of the National Academy of Sciences of the United States of America, www.pnas.org/cgi/doi/10.1073/pnas.1711842115

Castellacci, F., and Natera, J. (2013), The dynamics of national innovation systems: A panel co-integration analysis of the coevolution between innovative capability and absorptive capacity, Research Policy, 42, pp. 579–594.

Chaurasia, R. (2002), History of India, Earliest Times to 1200AD, Atlantic Publishers & Distributors (P), Ltd, Delhi.

Crossan, M., and Apaydin, M. (2010), A Multi-Dimensional Framework of Organisational Innovation: A Systematic Review of the Literature, of Management Studies, 47: 6.

Damanpour, F. (1992), Organisational Size and Innovation, Organisation Studies, 13/3, pp. 375–402.

Ettlie, J. (1983), Organisational Policy and Innovation among Suppliers to the Food Processing Sector, The Academy of Management Journal, Vol. 26, No. 1, pp. 27–44.

Goetzmann, W., and Koll, E. (2005), Paying in Paper, A Government Voucher from the Southern Song, IN The Origins of Value: The Financial Innovations That Created Modern Capital Markets, Oxford University Press, Oxford.

Guardian. (2018), Crayfish becomes online hero by detaching claw to escape boiling soup, https://www.theguardian.com/world/2018/jun/03/crayfish-becomes-online-hero-by-detaching-claw-to-escape-boiling-soup.

[38] Ettlie, J. (1983), Organisational Policy and Innovation among Suppliers to the Food Processing Sector, The Academy of Management Journal, Vol. 26, No. 1, pp. 27–44.

Haines, W. (1982), Psychoeconomics of Human Needs: Maslow's Hierarchy and Marshall's Organic Growth, Journal of Behavioral Economics, Vol. 11 (Winter), pp. 97–121.

Hamilton, G. (1996), The Organisational Foundations of Western and Chinese Commerce: A Historical and Comparative Analysis, IN Asian Business Networks, Hamilton, G. (Ed), Walter de Gruyter, Berlin.

Helfat, C. E., Finkelstein, S., Mitchell, W., Peteraf, M., Singh, H., Teece, D., and Winter, S. (2007). Dynamic Capabilities: Understanding Strategic Change in Organizations. New York: Wiley.

Kumari, A., and Bhaskar, S. (1998), Social Change Among Balijas, Majority Community of Andhra Pradesh, MD Publication PVT Ltd, New Delhi.

La Porta, R., Lopez-de-Silanes, F., Shleifer, A., and Vishny, R. (2000), Investor Protection and Corporate Governance, Journal of Financial Economics, 58, 3–27.

Nanjira, D. (2010), African Foreign Policy and Diplomacy from Antiquity to the 21st Century, Praeger, Santa Barbara, California.

Robinson, A. (2015), The Indus: Lost Civilisations, Reaktion Books Ltd, London.

Rosser, G. (2015), The Art of Solidarity in the Middle Ages: Guilds in England 1250–1550, Oxford University Press, Oxford.

Sapra, H., Subramanian, A., and Subramanian, K. (2014), Corporate Governance and Innovation: Theory and Evidence, Journal of Financial and Quantitative Analysis, Vol. 49, No. 4, pp. 957–1003.

2

Theory of Innovation and Causal Dynamics

The theories of technological change generally fall into two categories.[1] These include the Demand-pull and the Technology Push theories associated with technological change. Which of these two theories hold as implications not only for the determinants of how demand varies over time, but also for the determinants of innovation?[2] Moreover, changes in the level of innovative activity are a result of changes in demand, and are not the reason for the volatility of demand.[3] This knowledge is useful in pinpointing the actual causes of the changes in innovative activity.[4] The Demand-pull Theory of technical change suggests that technical change occurs due to market forces, customer needs, and the pricing mechanism. However, there are a number of weaknesses associated with Demand-pull Theory.[5] Firstly, technological change due to market conditions is seen as

[1] Dosi, G. (1982), Technological Paradigms and Technological Trajectories: A Suggested Interpretation of the Determinants and Directions of Technological Change, Research Policy 11, pp. 147–162.

[2] Geroski, P., and Walters, C. (1995), Innovative Activity and the Business Cycle, The Economic Journal, Vol. 105, No. 431, pp. 916–928.

[3] Ibid.

[4] Ibid.

[5] Ibid.

S. Ramesh, *The Rise of Empires*, https://doi.org/10.1007/978-3-030-01608-1_2

a passive and mechanical reaction.[6] Secondly, the theory is indeterministic about the reasons why technological changes take place as well as the timing of these occurrences.[7] Lastly, the theory does not explain why the inventive capacity of society changes over time, as well as its relationship to market conditions.[8] Moreover, in addition to these weaknesses, Demand-pull Theory has little empirical support in the majority of the studies conducted in the field.[9] On the other hand, Technology Push theory suggests that technical changes occur because of autonomous or semi-autonomous factors.[10] According to the Technology Push view, scientific research is the initiator of a sequence of stages, the end result of which is the innovation.[11] Thus, the literature places great significance on technological change as the primary driver of economic growth.[12,13] Innovation results in increased productivity, which results in increased performance.[14] The implication of this is that only the addition of creativity to economic life would result in economic development. Furthermore, studies have found that the supply side is more effective in generating innovation than is the demand side.[15] The entrepreneur is at the root of the supply side. The role of the entrepreneur is to combine existing ideas and resources in creative ways.[16] The entrepreneur is therefore the key facilitator of innovation. However, the strength of this may be affected by the strength of an

---

[6] Ibid.

[7] Ibid.

[8] Ibid.

[9] Ibid.

[10] Ibid.

[11] Saad, M. (2000), Development through Technology Transfer: Creating new Organisational and Cultural Understanding, Intellect Books, Portland, Oregon.

[12] Ayres, R.U. (1985), A Schumpeterian model of technological substitution. Technological Forecasting and Social Change 27: 375–383.

[13] Schmookler, J. (1966), Invention and economic growth, Harvard University Press, Cambridge, MA.

[14] Cainelli, G., Evangelista, R., and Savona, M. (2006), Innovation and Economic Performance in Services: A Firm Level Analysis, Cambridge Journal of Economics, 30, pp. 435–458.

[15] Knight, K.D. (1963), A Study of Technological Innovation. The Evolution of Digital Computers, DPhil Thesis, Carnegie, Institute of Technology, Pittsburgh.

[16] Schumpeter, J.A. (2000), Entrepreneurship as Innovation. In Swedberg, R. (Ed), Entrepreneurship, Oxford University Press, Oxford pp. 51–75.

individual's relational context, for example, the extent of the development of network ties in the context of diversity, strength, and content.[17] The weakness of the Technology Push Theory is that it fails to consider the importance of economic factors, as well as changes in economic factors in affecting the direction of the innovation process.[18] In this case, it is easy to see that there is a causal relationship between economic development and stability with the direction of technological change.[19] Therefore, in order to be of relevance to the real world, it is imperative that any theory of technological change and innovation is able to explain the interactive mechanisms between economic development and stability and the direction of technological change.[20] Moreover, while neoclassical economics considers the economy from the perspective of general equilibrium, economic change is a dynamic process that is interactive with innovation. Furthermore, similarly to the situation in which some firms perform financially better than others due to a resource advantage,[21] the same may also be said of states. The resource reserves and the resource capacity of states change over time, and this may explain the rise and fall of states, why civilisations endure and why they may not. Nevertheless, from one perspective, it can be argued that differences in technology between contemporary states and civilisations is due to differences in the levels of knowledge accumulation rather than because of differences in resource endowments.[22] Unfortunately, the existing theories of technical change are unable to decipher what these interactive mechanisms are, how they arise, and how they behave. However, there may be a link between the

[17] Ruef, M. (2002), Strong ties, weak ties and islands: Structural and cultural predictors of organisational innovation, Industrial and Corporate Change, Vol. 11, No. 3, pp. 427–449.

[18] Dosi, G. (1982), Technological Paradigms and Technological Trajectories: A Suggested Interpretation of the Determinants and Directions of Technological Change, Research Policy 11, pp. 147–162.

[19] Ibid.

[20] Ibid.

[21] Hunt, S., and Morgan, R. (1996), The Resource-Advantage Theory of Competition: Dynamics, Path Dependencies, and Evolutionary Dimensions, Journal of Marketing, Vol. 60, No. 4, pp. 107–114.

[22] Ericsson, J., and Irandoust, M. (2001), On the Causality Between Foreign Direct Investment and Output: A Comparative Study, The International Trade Journal, Volume XV, No. 1.

innovation process and the interaction between institutional and economic factors.[23] For example, stable government and the rule of law may provide economic stability. This could be reflected by a sustainable food source, more leisure time, more time for learning, more engagement with cognition, leading to greater creativity, and innovation. Furthermore, stable government and economic stability will allow for the longevity of the state, allowing for the accumulation of knowledge. Moreover, state ideology and support would be able to support creativity and innovation, and specific needs would foster specific innovations, for example, agriculture and irrigation, the military and the chariot, defence and city walls. As the state grows, economic and social activities would become more and more hierarchically organised. This would lead to an increased division of labour. Specific tasks would be performed by specific individuals. As a result of performing tasks repetitively, better ways of doing the same thing would be recognised and innovation would result. Therefore, from this line of reasoning, it is clear that a strong state would favour greater hierarchical organisation over time, and this in turn would give rise to the development of complex technological systems.[24] However, when a state becomes unstable, resources become concentrated in its defence, and power structures become thinly spread, allowing for a decline in innovation.[25] This may be because the level of information in society as well as the level of group communication is reduced. It has been found that both of these communication variables are more important as the drivers of organisational innovation than are motivational variables such as perceptions of equity, social pressures, and expectations of benefits from carrying out a task.[26]

[23] Dosi, G. (1982), Technological Paradigms and Technological Trajectories: A Suggested Interpretation of the Determinants and Directions of Technological Change, Research Policy 11, pp. 147–162.

[24] Murmann, J., and Frenken, K. (2006), Toward a Systematic Framework for Research on Dominant Designs, Technological Innovations, and Industrial Change, Research Policy, 35, pp. 925–952.

[25] Mone, M., McKinley, W., and Barker, V. (1998), Organisational Decline and Innovation: A Contingency Framework, The Academy of Management Review, Vol. 23, No. 1, pp. 115–132.

[26] Monge, P., Cozzens, M., and Contractor, N. (1992), Communication and Motivational Predictors of the Dynamics of Organisational In novation, Organisation Science, Vol. 3, No. 2.

Analysis of the Innovation Process

The process of innovation can be defined in terms of its result, the contemporary use of an idea or behaviour which is new to society.[27] However, the process of innovation may also involve improving a pre-existing idea or using it for a new purpose.[28] The process of creativity is closely allied to the process of innovation. In this case, creativity can be defined as a new way of thinking that allows one to transcend contemporary thought so that the result is something new and tangible, either a behaviour, good, work of art, process, or even a political/economic policy. Some people are more creative than others. In this case, the key drivers of innovation are people and their experiences, because ultimately, knowledge is owned by people.[29] However, the link between the factors that bring about organisational change through innovation is known as causal agency. For spontaneous innovation, organisational change may not be important. However, the emergence of organisational capacity, its development and growth over time, and its ending does impact on innovation as a process.[30] Causal systems are defined by the level of dynamism as well as by the level of interaction between the causal factors.[31] Low dimensional causal systems are characterised by periodic and chaotic dynamics.[32] On the other hand, high dimensional causal systems are characterised by white and pink noise dynamics.[33] In the case of periodic and white noise dynamics, there is a linear relationship between the causal agents, such that they act independently.[34] But in the case of chaotic

[27] Hage, J.T. (1999), Organisational Innovation and Organisational Change, Annual Reviews Sociology, Vol. 25, pp. 597–622.

[28] Woodman, R., Sawyer, J., and Griffin, R. (1993), Toward a Theory of Organisational Creativity, The Academy of Management Review, Vol. 18, No. 2, pp. 293–321.

[29] Harkema, S. (2003) 'A complex adaptive perspective on learning within innovation projects', The LearningOrganization,Vol.10Issue:6,pp.340–346,https://doi.org/10.1108/09696470310497177.

[30] Van de Ven, A., and Huber, G. (1990), Longitudinal Field Research Methods for Studying Processes of Organisational Change, Organisation Science, Vol. 1, No. 3, pp. 213–219.

[31] Dooley, K., and Ven, A. (1999), Explaining Complex Organisational Dynamics, Organisational Science, Vol. 10, No. 3.

[32] Ibid.

[33] Ibid.

[34] Dooley, K., and Ven, A. (1999), Explaining Complex Organisational Dynamics, Organisational Science, Vol. 10, No. 3.

and pink dynamics, there is a non-linear relationship between the causal agents, suggesting that the causal factors act interdependently.[35] Knowledge is acquired by using one's cognition in order to perceive the real world, and then to innovate. This type of causal agency between people and innovation, bringing about social change, is known as the organisational imperative.[36] On the other hand, the technological imperative asserts that technology is the causal agent bringing about social change.[37] However, a middle-of-the-way perspective, the emergent imperative, asserts that the causal agent of social change is the complex series of interactions between technology and people.[38] This seems to be a more plausible cause of causal change in which human actors innovate using their cognitive capacity, reflected by technological innovations which facilitate organisational change in society. For example, the start of the industrial revolution in the latter part of the eighteenth century in England brought about technological changes that, over a period of time, increased productive capacity, as well as increasing the size of urban centres. The increased productive capacity meant that there was now more to trade. This facilitated the formulation of a new national economic policy, Mercantilism, which would eventually lead to the formation of the British Empire. A 'state' becomes an 'empire' when it has a superior form of political and firm level governance over other contemporary states, which allows it to more efficiently organise and allocate its economic, physical, political, and intellectual resources in order acquire the land, labour, and material wealth of militarily and economically weaker states. Great Britain's evolving system of governance, which was superior to that of other states in the context of the efficient organisation and allocation of resources, from the twelfth century onwards, coupled with its embracement of Mercantilism from the sixteenth century, facilitated its emergence as an 'empire' over the next 450 years. In essence, Mercantilism was the radical change of a society and its institutions such that national institutions were replaced by trans-national ones, under the

[35] Ibid.

[36] Markus, M., and Robey, D. (1988), Informational Technology and Organisational Change: Causal Structure in Theory and Research, Management Science, Vol. 34, No. 5, pp. 583–598.

[37] Ibid.

[38] Ibid.

umbrella of a dominant nation state controlling trade with its constituent parts.[39] Besides Great Britain, this policy was also followed by Prussia, leading to the unification of the German states into a single country in 1871.[40]

The real world will provide more information than can be processed by human cognition.[41] The latter, therefore, acts as a filter to process the information provided in its best representable, most concise form.[42] Nevertheless, the way in which the innovation process takes place can be analysed from the perspective of both Actor-Network Theory (ANT) and Cultural-Historical Activity Theory (AT).[43] ANT suggests that the innovation process should be studied by analysing the development of the innovation as well as the network of agents connected to it, simultaneously.[44] At a wider level, an innovation system has structural components such as agents, networks and institutions in order to perform a range of functions.[45] These include the development of knowledge, the mobilisation of resources, the state of development of market forces, the power of the state, the state's impact on the direction of innovation, the development of entrepreneurship, and the extent of external economies.[46] However, in order for the structural components of an innovation system to be able to jointly carry out the required functions, there is a high level of dependency on the development of people. Nevertheless, people develop in a cultural context.[47] Therefore, in this case, AT suggests that

[39] Schmoller, G. (1884), The Mercantile System and Its Historical Significance.

[40] Carr, W. (2013), The Origins of the Wars of German Unification, Routledge, London.

[41] McGuire, W.J. (1983), Search for the Self: Going beyond self-esteem and the reactive self. In Zucker, R., Arnoff, J., and Rabin, A. (Eds), Personality and the Prediction of Behaviour, Academic Press, New York.

[42] McGuire, W. (1983), A Contextualist Theory of Knowledge: Its Implications For Innovation and Reform in Psychological Research, Advances in Experimental Social Psychology, Vol. 16.

[43] Miettinen, R. (1999), The riddle of things: Activity theory and actor-network theory as approaches to studying innovations, Mind, Culture, and Activity, 6:3, 170–195, DOI:https://doi.org/10.1080/10749039909524725

[44] Ibid.

[45] Bergek, A., Jacobson, S., Carlsson, B., Lindmark, S., and Rickne, A. (2008), Analysing the functional dynamics of technological innovation systems: A Scheme of Analysis, Research Policy, 37, pp. 407–429.

[46] Ibid.

[47] Oers, B., Wardekker, W., Elbers, E., and Veer, R. (2008), The Transformation of Learning: Advances in Cultural-Historical Activity Theory, Cambridge University Press, Cambridge.

the innovation process should be analysed and studied from a cultural-historical perspective. This is specifically because the cultural practices of a community as well as its immediate environment change over time.[48] Moreover, there is a great deal of variation in the behaviours of people with regards to different cultural communities as well as different time periods.[49] There is significant empirical and theoretical support for this assertion.[50] The implication of this assertion is that the variations with people behave over time may be affected by learning leading to innovations. For example, the innovations of electricity and the light bulb allowed people to work longer hours. The invention of the steam engine in the late eighteenth century was driven by socio-political changes. In this context, there was an emphasis on Mercantilism, entrepreneurship, property rights, and less government interference in the market at this time. Furthermore, there was a diffusion of knowledge from one society to another because of written and printed texts. Accumulated historical innovations associated with shipping (ship building, the compass) and the military (organisation of people, gunpowder) allowed for territorial expansion. This facilitated the culture and the history of one culture impacting on that of another. For example, after the fifteenth century, European nations began to explore and colonise other parts of the world already inhabited by indigenous peoples. Thus, socio-political changes and accumulated knowledge facilitated historical transformation. This was probably the root of the rise of the British Empire from the seventeenth century onwards. However, historical transformation could only have occurred because of a learning process. According to the ancient Greek philosopher, this learning process involved learning by doing.[51] In other words, when an individual conscientiously noted a new way of doing something, then, this knowledge would diffuse throughout the

[48] Ibid.

[49] Rogoff, B. (2003), The Cultural Nature of Human Development, Oxford University Press, Oxford.

[50] Oers, B., Wardekker, W., Elbers, E., and Veer, R. (2008), The Transformation of Learning: Advances in Cultural-Historical Activity Theory, Cambridge University Press, Cambridge.

[51] Oers, B., Wardekker, W., Elbers, E., and Veer, R. (2008), The Transformation of Learning: Advances in Cultural-Historical Activity Theory, Cambridge University Press, Cambridge.

individual's society, and would eventually be improved upon by others. However, in general, learning facilitates the development of creative abilities, and thus, one's ability to generate new ideas. In this way, learning brings about a permanent change in the way in which people behave, and this occurs entirely because of the practice of learning.[52] But the practice of learning depends upon human cognition, and the development of human cognition over time is affected by a range of socio (cultural practice, beliefs, behaviours) and economic (income levels, distribution of income) factors. Therefore, it is important that the causal and dynamic relationships between the various socio and economic factors are jointly analysed because of their tendency to co-evolve.[53] Therefore, because of the importance of human cognition to innovation, it is important not to analyse socioeconomic factors in isolation.[54]

Cognition

Human cognition involves the accumulation of information, either through the senses (ear, nose, eyes) as well as through contact with other individuals, such that human behaviour is optimised to the environment.[55] In other words, there are cognitive factors common to us all that may result in the same response to similar environmental factors. For example, if someone touches a hot object, sensory perception and cognitive processes lead the person to automatically withdraw the hand. Similarly, if the weather is extremely cold, we wear as much warm clothing as possible when we go out. And, if someone is unpleasant to us, we avoid their company and seek out the company of those others whose company we enjoy. The optimisation of human behaviour to suit the environment is like the concept of rationality, which is assumed of consumer behaviour in

---

[52] Saltz, E. (1971), The Cognition bases of human learning, Homewood, Dorsey Press.

[53] Room, G., Spence, A., Evangelou, E., Zeppini, P., Pugliese, E., Napolitano, L. (2017), Dynamics of Cumulative Innovation in Complex Social Systems (DCICSS). UK: University of Bath.

[54] Ibid.

[55] Anderson, J. (1991), Is human cognition adaptive, Behavioural and Brain Sciences, 14, pp. 471–517.

Economics. However, the optimisation of human behaviour is less specific than the economic assumption of rationality in consumer choice. Human cognition, therefore, facilitates the formulation of theories and hypotheses as a result of the observation of events, either directly or through a process of trial and error. For example, in sixteenth-century England, the merchant-turned-political economist Thomas Mun, was able to draw on his experiences as a merchant in Italy and the Middle East, to support the policy of Mercantilism in a political, theoretical, and philosophical manner. According to one perspective, innovation or a disruptive theory arises from a two-stage process.[56] The two stages, in successive order, are the descriptive stage and the normative stage.[57] While each stage involves a process of the three steps of observation, categorisation, and the definition of relationships.[58] Observation, involving the senses, gives rise to questions in the mind as to why something is happening. Categorisation involves de-complexing a real-world observation using cognitive capacity into a framework of consequential relationships between variables. For example, in early human history, early humans may have observed a lightning strike on forests or savannahs and seen the resulting flames. The cognitive capacity of early humans would then have been engaged in trying to establish a link between the lightning and the resulting fire. Perhaps, through a process of trial and error, as well as luck, the early humans would have determined that there was a consequential link between the lightning and dry grass and wood. The third step would be to define the relationship between high temperatures, caused by the lightning, and dry grass and wood, perhaps by proving it. The normative aspect of theory formation is, however, more practical, and involves overcoming any anomalies which may not 'fit in' with the hypothesis—lightning, causes high temperatures, and combusts dry grass or dry wood. The normative aspect of theory formation also involves the three stages of observation, categorisation, and relationship definition.[59] It is at the latter stage of normative theory evolution that disruptive innovations—in this case, the ability of early humans to make

[56] Christensen, C. (2006), The Ongoing Process of Building a Theory of Disruption, J. Prod. Innov. Manag, 23, pp. 39–55.

[57] Ibid.

[58] Ibid.

[59] Ibid.

fire—emerge. The knowledge of fire making would then have diffused through ancient human society, perhaps by conflict and/or co-operation, but definitely because contact between groups was possible due to limited distances.[60,61] Knowledge production would have been more efficient when the human occupation of land was dense compared to it being less dense.[62] It may have been the case that youngsters within the group learn from the adults, as well as members of the group learning from each other. In humanity's closest relative, the chimpanzee, it has been observed that infants are able to use stone tools as hammers to break open oil palm nuts.[63] They must have been able to achieve this feat through a process. Firstly, by observing another member of the group, perhaps a parent. Second, cognitively putting together the relationship between the stone tool and the oil palm nut. And lastly, putting the relationship into action by using a stone tool to crack the oil palm nut themselves. More formally, the diffusion of innovations in a wider context can occur through direct communication,[64] equivalent roles, and performing the same tasks in different locations,[65] institutional needs,[66] and network availability.[67] In the case of Homo Sapiens, then, learning would have been more effective in smaller groups, the modern-day equivalent of the clustering of firms and agglomeration effects.[68] However, the process of knowledge diffusion

[60] Feldman, M.P. (1994a). The Geography of Innovation. Kluwer Academic, Boston.

[61] Feldman, M. (1994b). 'Knowledge complementarity and innovation'. Small Business Economics 6 (3), 363–372.

[62] Audretsch, D., and Feldman, M. (2004), Knowledge Spillovers and the Geography of Innovation, IN Handbook of Regional and Urban Economics, Volume 4, Henderson, J., and Thisse, J. (Eds), Elsevier B.V.

[63] Inoue-Nakamura, N., and Matsuzawa, T. (1997), Development of Stone Tool Use by Wild Chimpanzees, Journal of Comparative Psychology, Vol. III (2), pp. 159–173.

[64] Coleman, J.S., Katz, E., and Menzel, H. (1966), Medical Innovation, Bobbs-Merrill, Indianapolis.

[65] Burt, R. (1987), 'Social Contagion and Innovation: Cohesion versus Structural Equivalence', American Journal of Sociology, 92, 1287–1335.

[66] Strang, D., and Meyer, J. (1993), Institutional Conditions for Diffusion, Theory and Society, 22, pp. 487–511.

[67] Hedstrom, P., Sandell, R., and Stern, C. (2000), 'Mesolevel networks and the diffusion of social movements: the case of the Swedish Social Democratic Party', American Journal of Sociology, 106, pp. 145–172.

[68] Malmberg, A., and Maskell, P. (2002), The Elusive Concept of Localisation Economies: Towards a Knowledge-Based Theory of Spatial Clustering, Environment and Planning, Vol. 34, pp. 429–449.

between individuals is very different from that between firms.[69] Moreover, there is controversy as to whether the level of innovation is higher in smaller or larger firms.[70]

Central to innovation and behaviour is human cognition. Advances in the study of human cognition have proceeded in three stages.[71] The first stage occurred in the 1950s and the 1960s—a time during which human cognition was studied using traditional psychophysics and experimental psychology. In the second stage, starting in the 1970s, human cognition was studied using emerging computational analysis. In the third stage, from the mid-1980s, neuropsychology and animal neurophysiology contributed evidence to the study of human cognition. In more recent times, human cognition has been studied by scanning the brain to analyse activity in specific areas.[72] However, the problem with the study of human cognition is that it has taken place under controlled situations in a laboratory. The results of such analysis are abstract, especially in the context that cognitive processes depend on both environmental factors and bodily metabolism.[73] Therefore, the best approach to study human cognition is to gather insights into human behaviour and cognition from outcomes in real-world situations. This is because the objective of the study of human cognition is to better understand how human cognition functions in complex real-world circumstances.[74] If this could be accurately done, then an element of predictability would be introduced into human cognition and behaviour. The implications of this is that governments may be able to design social policies based on the results, if they are generalisable, of the studies of human cognition and behaviour. However, there

[69] Wolfe, R. (1994), Organisational Innovation: Review, Critique and Suggested Research Directions, Journal of Management Studies, 31:3.

[70] Nooteboom, B. (1994), Innovation and Diffusion in Small Firms: Theory & Evidence, Small Business Economics, 6, pp. 327–347.

[71] Van Kleeck, M. H., & Kosslyn, S. M. (1991), Visual information processing: A perspective, IN: Attention and performance XIV, Meyer, D., & Kornblum, S. (Eds.), Bradford Books/MIT Press, Cambridge, MA.

[72] Kingstone, A., Smilek, D., and Eastwood, J. (2008), Cognitive Ethology: A New Approach for Studying Human Cognition, British Journal of Psychology, 99, pp. 317–340.

[73] Ibid.

[74] Ibid.

are many problems regarding how such real-world studies can be conducted.[75] Firstly, because no controls will be in place, there will be variability in the results. For example, many people may behave differently to the same environmental factors at the same time.[76] On the other hand, the same person may behave differently to the same environmental factors at different times.[77] Secondly, there is no preconceived model or methodology upon which such studies could be conducted. This is more so with a causal mechanism-based approach.[78] In this context, researchers have tended to favour a non-personal approach to understand the brain mechanisms which form the foundation of cognition.[79] However, such non-personal approaches may fall short of delivering a valid portrayal of how human cognition functions in complex real-world situations, for many reasons.[80] Firstly, there is no contribution to understanding what cognitive performance is. Secondly, there is no indication as to the correlation between cognitive performance and cognitive systems. Lastly, non-personal studies are not able to contribute as to how cognition functions when many individuals are involved in solving a common problem. However, a causal mechanism-based analysis could be conducted to evaluate such macro-level innovation in the context of situational mechanisms, action formation mechanisms, and transformational mechanisms.[81] Situational mechanisms refer to the factors in the macro environment which gives rise to action formation mechanisms. This reflects the beliefs, goals, and opportunities that impact on individual behaviour.[82] The transformational mechanism reflects how the behaviour of many can

[75] Ibid.

[76] Ibid.

[77] Ibid.

[78] Hedstrom, P., and Wennberg, K. (2017), Causal mechanism in Organisation and Innovation Studies, Innovation, 19:1, pp. 91–102.

[79] Kingstone, A., Smilek, D., and Eastwood, J. (2008), Cognitive Ethology: A New Approach for Studying Human Cognition, British Journal of Psychology, 99, pp. 317–340.

[80] Ibid.

[81] Hedstrom, P., and Wennberg, K. (2017), Causal mechanism in Organisation and Innovation Studies, Innovation, 19:1, pp. 91–102.

[82] Hedstrom, P., and Wennberg, K. (2017), Causal mechanism in Organisation and Innovation Studies, Innovation, 19:1, pp. 91–102.

result in unintended macro-level outcomes.[83] However, the problem with investigating the link between micro-level and macro-level innovation processes is the lack of suitable methodology,[84] although there has been some methodological development with the use of mathematical models and the emerging big data. Therefore, in the light of these short comings of the non-personal approach to understanding human cognition, researchers should favour a study of a personal approach to understanding human cognition in the context of how people interact with their environment.[85] Nevertheless, based on the principle that every individual is based on the same biological template, it should be possible to objectively identify a behavioural and cognitive template that can be used to forecast the random responses of individuals to environmental factors, the effects of which have already been recorded through observation on other individuals.

The basis for understanding creativity and innovation could depend on a better understanding of human cognition.[86]While creativity is the basis for innovation, as a process, it involves creating original solutions to novel problems, ranging from technology to policy.[87] However, this depends upon four important steps, involving expertise, knowledge, aptitude, and strategies.[88] Aptitude can be best described in the context of four processes associated with creative thinking. These include the definition of the problem, the gathering of information and data, selecting an appropriate concept, and then combining concepts.[89] However, human capacity and the potential to innovate have to be separated from the economic

---

[83] Ibid.

[84] Coleman, J. S, (1986), Social theory, social research, and a theory of action, American Journal of Sociology, *91*, 1309–1335.

[85] Ibid.

[86] Mumford, M., Hunter, S., and Byrne, C. (2009), What is the Fundamental? The Role of Cognition in Creativity and Innovation, Industrial and Organisational Psychology 2, pp. 353–356.

[87] Mumford, M. D., & Gustafson, S. B. (2007), Creative thought: Cognition and problem solving in a dynamic system. IN Creativity research handbook (Vol. 2 pp. 33–77), M. A. Runco (Ed.), Cresskill, NJ: Hampton.

[88] Mumford, M., Hunter, S., and Byrne, C. (2009), What is the Fundamental? The Role of Cognition in Creativity and Innovation, Industrial and Organisational Psychology 2, pp. 353–356.

[89] Ibid.

background. This is because, without a conducive economic environment, the would-be innovator would not have the financial resources to innovate 'big'. The emergent Mercantilist economy of the sixteenth century onwards created an affluent merchant class with the financial resources to subsidise the livelihoods of inventors. This was the pathway for the rapid technological development which took place at the end of the eighteenth century because of the increasing costs of production faced by the merchant class of the time. To mitigate the rising costs of production, the merchants funded innovation, either by themselves or by others. For example, Arkwright developed upon Hargreaves' Spinning Jenny and invented a cotton spinning wheel which produced better quality cotton yarn, although the patent for the machine was financed by others.[90] The patent for Arkwright's cotton spinning machine was granted in 1769,[91] as was his patent for the source of power of the machines, the water frame.[92] These innovations gave England a competitive advantage in the production of quality cotton yarn as well as allowing Arkwright to develop his system of cotton factories, representing a new innovation in industrial organisation.[93] The cheaper, quality cloth took the place of Indian textiles in trade with Africa, as well as triggering new innovations such as steam power, and new industries such as the manufacture of machinery and machine tools.[94] In order to incentivise innovation and to foster technological diffusion from continental Europe, sixteenth- century England adopted a patent system which allowed inventors to move to England, where their innovations were given legal protection. However, this was with the caveat that English workers and English industry would benefit from such innovation.[95] Nevertheless, significant English innovations

[90] Hills, R. (1970), Sir Richard Arkwright and his patent granted in 1769, The Royal Society Journal of the History of Science, Vol. 24, Issue 2.

[91] Ibid.

[92] Hill, D. (1969), Richard Arkwright and the Water Frame, Proceedings of the Institution of Mechanical Engineers, Vol. 184, Issue 1.

[93] Hills, R. (1970), Sir Richard Arkwright and his patent granted in 1769, The Royal Society Journal of the History of Science, Vol. 24, Issue 2.

[94] Ibid.

[95] Macleod, C. (1988), Inventing the Industrial Revolution, The English Patent System, 1660–1800, Cambridge University Press, Cambridge.

also occurred. For example, James Watt re-invented the steam engine with the financial support of Matthew Boulton, a Birmingham merchant.[96] Watt was called upon to repair a Newcomen-type atmospheric steam engine in 1763, at the University of Glasgow. However, the problem was that, to cool the cylinder, cold water had to be drawn into it, with the effect that the cylinder was continuously being cooled and heated, leading to the inefficient and expensive use of coal.[97] Watt realised the inefficiency of the design due to his superior knowledge of mathematics, in comparison to Newcomen. Watts solution was to add another chamber, the condenser in which the hot water could condense, denigrating the need to put cold water into the heated cylinder, which would remain hot until more steam arrived.[98] While Watt patented the steam engine in 1769, it did not finally go into production, with Boulton's help, until 1776.[99] However, when production did begin, existing craftsman had to 'skill-up' through on-the-job training, rather than through a scholarly approach.[100] Moreover, techniques also had to be scaled up, a shift from making scientific instruments to steam engines, as well as advances in precision measurement.[101] The steam engine facilitated the spatial reorganisation of industrial production, such that it had no longer to be located near sources of water.

The creative process of discovery and innovation can also be thought of as a system.[102] This system involves the interaction of the sociocultural environment with the factors of cognition such as intelligence, information retention, individual characteristics, the desire to be creative, and learning ability.[103] Learning ability in innovative projects may not simply

---

[96] Scherer, F.M. (1965), Invention and Innovation in the Watt-Boulton Steam-Engine Venture, Technology & Culture, Vol. 6, No. 2, pp. 165–187.

[97] Collier, J. (2006), Great Inventions – The Steam Engine, Marshall Cavendish Benchmark, New York.

[98] Ibid.

[99] Ibid.

[100] Russell, B. (2014), James Watt: making the World Anew, Reaktion Books Ltd., London.

[101] Ibid.

[102] Findlay, C., and Lumsden, C. (1988), The Creative Mind: Toward an Evolutionary Theory of Discovery and Innovation, J. Social Biol. Struct, 11, pp. 3–55.

[103] Ibid.

be based on a linear sender-receiver model, but should be viewed as being a more effective process, the greater the level of interaction within a team of people.[104] The conceptualisation of the creative process as being an integrated system between the sociocultural environment and individual cognition implies that its hierarchy cannot be reduced to a single entity.[105] Furthermore, while individual cognition is a biological phenomenon, it develops over time, through the passing of genetic material over generations. The transmission of this genetic material is affected by the sociocultural environment associated with each generation.[106] Moreover, the creative process may result from the linking of networks in the brain that have resulted from previous episodes of creative need.[107] Specific environmental stimuli may stimulate the body to adapt through the biological pathways of genetic change by allowing the DNA and the RNA in cells to produce event-related proteins that may facilitate neuron development and connection in the brain.[108] Such specific environmental stimuli may be related to educational activities. In this case, learning may trigger the formulation of ideas, which in turn stimulates the development of new neurons and connections in the brain. Originally, it had been thought that neurons developed in the brain only during the embryonic and infant stages of human development.[109] However, it has recently been found that immature neurons in the nasal region and stem cells in the brain may develop into neurons and migrate to the olfactory bulb of the brain, where they transform into fully functioning glia or neurons.[110] But it would seem that the neurons connect glia regions in the brain, rather than the glia just being structural elements to active neurons.[111] In

[104] Harkema, s. (2003), A complex adaptive perspective on learning within innovation projects, The LearningOrganization,Vol.10Issue:6,pp.340–346,https://doi.org/10.1108/09696470310497177.

[105] Findlay, C., and Lumsden, C. (1988), The Creative Mind: Toward an Evolutionary Theory of Discovery and Innovation, J. Social Biol. Struct, 11, pp. 3–55.

[106] Ibid.

[107] Ibid.

[108] Ibid.

[109] Kalat, J. (2007), Biological Psychology, Thomson Wadsworth, Belmont, California, USA.

[110] Ibid.

[111] Koob, A. (2009), The Root of Thought: Unlocking Glia – The Brain Cell That Will Help Us Sharpen Our Wits, Heal Injury and Treat Brain Disease, Pearson Education, New Jersey, USA.

this context, the role of neurons in intelligence and creativity is being actively challenged.[112] New evidence suggests that the glia perform two functions in brain physiology.[113] Firstly, the glia can replace themselves as well as neurons acting as stem cells. Secondly, the glia may act as stores of information. It has also been found that higher intelligence requires a higher ratio of glia to neurons. The 'firing' of the neurons may therefore "connect" different regional glia, allowing for different information stores to be accessed simultaneously and thoughts to form. Experiences and observations will be connected through a thought process, allowing insights to form, with creativity and innovation following. Therefore, intentional actions, such as learning, may be best at directing the pace of technological change in some phases of innovation than in others.[114] But, whatever the case, the innovation will change both innovated system as well as the innovator.[115] However, creativity and innovation does not have to result in physical solutions, but also intangible ones such as a specific economic policy or doctrine. The emergence of Mercantilism giving rise to the British Empire is one such example.

Innovation of Thought: Mercantilism

Before, the fifteenth century, the economy of England resembled its natural state, essentially a subsistence economy in which people tilled the land in order to feed themselves and pay the lords of the land, taxes.[116] There was little surplus agricultural goods left that could be exchanged for the goods produced by the merchants, who were organised into

[112] Ibid.

[113] Koob, A. (2009), The Root of Thought: Unlocking Glia – The Brain Cell That Will Help Us Sharpen Our Wits, Heal Injury and Treat Brain Disease, Pearson Education, New Jersey, USA.

[114] Hughes, T.P. (1983), Networks of Power: Electrification of Western Society 1880–1930, Johns Hopkins Univ. Press, Baltimore, MD.

[115] Etzkowitz, H., and Leydesdorff, L. (2000), The Dynamics of Innovation; From National Systems and 'Mode 2' to a Triple Helix of University-industry-government relations, Research Policy, 29, pp. 109–123.

[116] Gaido, D. (2016), Rudolf Hilferding on English Mercantilism, History of Political Economy, 48:3.

monopolies known as guilds by trade, who lived in the towns as well as in the cities.[117] The guilds placed constraints on the levels of employment of workers, as well as on the extent of production, in order to ensure that prices were high.[118] However, this system began to break down in the sixteenth and the sixteenth centuries, with the emergence and development of the monetary economy, as well as the rise of the merchant class.[119] There was a shift in traditional trading patterns from the Levant towards America and India.[120] Specific state-chartered companies were formed to manage trade with these regions and, as a result, colony formation and the plunder of distant civilisations by the European powers followed.[121] The result was an accumulation, by the merchant class, of precious metals such as gold and silver.[122] The rising wealth of the merchants allowed them to escape the power of the guilds, and to increase production using the 'putting out system'. This involved the merchants transporting raw materials to households in rural areas and the households producing goods such as clothes and shoes for the merchants to sell in the cities. As a result, there was a shift away from subsistence farming in the rural economy towards manual production. The doctrine of Mercantilism then emerged against the backdrop of the transition of the economy from a feudal one to a primordial capitalist economy.[123] However, trade in medieval England was tolerated if it did not impact on the social position of those individuals involved in trade.[124] But the expansion of foreign trade resulted in a conflict between the secular pursuits of the merchants and the non-secular interests of the state.[125] However, not only were the merchants accumulating wealth, but also increasingly becoming active in

[117] Ibid.
[118] Ibid.
[119] Ibid.
[120] Ibid.
[121] Ibid.
[122] Ibid.
[123] Ibid.
[124] Elmslie, B. (2015), Early English Mercantilists and the Support of Liberal Institutions, History of Political Economy, 47:3.
[125] Ibid.

politics and parliament. This served to counter-balance the hostility of the English Crown towards trade in a positive manner.[126] Furthermore, Mercantilist writers were devising economic theories such as opportunity cost, interest for delayed consumption, and a circular flow of income to support their argument in favour of trade.[127] Writers of the time, such as Thomas Mun, argued that the natural wealth of England could only be used profitably to cure the English of their idleness through increased trade, supported by a strengthened parliament and property rights.[128] English capital was being depleted due to the consumption of luxury goods and the maintenance of the poor, which composed a significant part of society.[129] The only way in which the haemorrhaging of English capital could be stopped was by the export of English goods to overseas markets, which would earn England financial resources in terms of gold and silver reserves.

It was from the medieval merchants of fifteenth-century England that the early institutions upon which Capitalism was to be based, began to form.[130] Moreover, it was also at this time that the institutions required for the taking off of England's Industrial Revolution were being formed.[131] In this context, the Mercantilist system allowed for the support of economic activities associated with commerce and industry, which facilitated the accumulation of capital and allowed the transformation of society from a feudal state to a bourgeois society.[132] Smith, therefore, viewed Mercantilism has a clash between the land-owning classes and the

---

[126] Ibid.

[127] Irwin, D. (1996), Against the Tide: An Intellectual History of Free Trade. Princeton, N.J.: Princeton University Press.

[128] Mun, T. (1621) 1986, A Discourse of Trade, from England unto the East-Indies, New York: Augustus M. Kelley.

[129] Appleby, J. (1976), Ideology and Theory: The Tension Between Political and Economic Liberalism in Seventeenth Century England, The American Historical Review, Vol. 81, No. 3, pp. 499–515.

[130] Elmslie, B. (2015), Early English Mercantilists and the Support of Liberal Institutions, History of Political Economy, 47:3.

[131] Ibid.

[132] Herlitz, L. (1964) The concept of mercantilism, Scandinavian Economic History Review, 12:2, 101–120, DOI: https://doi.org/10.1080/03585522.1964.10407639.

merchants.[133] However, while Smith held the view that Mercantilism led to economic inefficiency, it was David Ricardo and his Law of Comparative Advantage that led to the notion that, in the case of free trade, the balance of payments would take care of itself.[134] Ricardo's theory led to the British Empire embracing free trade and to reject Mercantilist tenets. Moreover, the time had come to allow other countries to trade freely with Great Britain and its colonies.[135] The momentum for this gathered pace towards 1846, and the repeal of the Corn Laws.[136] However, at around the same time, Friedrich Liste was attempting a resurgence of Mercantilism. In this context, Liste asserted that only countries in the 'temperate zone' (Europe) should industrialise by exploiting the natural resources of countries in the 'torrid zone', such as India, by dominating it.[137]

As Mercantilism emerged in England in 1713, political power was in the hands of the land owners and the merchant manufacturers.[138] As Mercantilism was given government endorsement, it seems clear that the merchant manufacturers were the dominant political force. Moreover, at the time, monopoly-type companies engaging in English trade were eliminated, except perhaps in the context of trade with the East Indies.[139] Mercantilists were opposed to monopoly practices.[140] Mercantilism, it would seem, was a doctrine that was state led. But mercantilism itself emphasised a strong state.[141] A strong state could balance the levy-charging power of the lords and the increasing economic autonomy of the towns in

[133] Ibid.

[134] Blaug, M. (1964), Economic Theory and Economic History in Great Britain, 1650–1776, Past & Present, No. 28, pp. 111–116.

[135] Mirza, R. (2007), The Rise and Fall of the American Empire: A Re-Interpretation of History, Economics and Philosophy: 1492–2006, Trafford Publishing, Victoria, Canada.

[136] Ibid.

[137] Pradella, L. (2014), New Developments and the Origins of Methodological Nationalism, Competition and Change, Vol. 18, No. 2, pp. 180–193.

[138] Coleman, D. (1980), Mercantilism Revisited, The Historical Journal, Vol. 23, No. 4, pp. 773–791.

[139] Ibid.

[140] Grampp, W. (1952), The Liberal Elements in English Mercantilism, The Quarterly Journal of Economics, Vol. 66, No. 4, pp. 465–501.

[141] Johnson, E.A.J. (1928), Some Evidence of Mercantilism in the Massachusetts-Bay, The New England Quarterly, Vol. 1, No. 3, pp. 371–395.

which the merchant guilds operated.[142] This could be achieved in several ways, which included the accumulation of gold and silver, denigrating the power of the guilds, increasing the population, developing a strong manufacturing industry, as well as enacting navigation laws.[143] However, although local monopolies had been denigrated and brought under state control by the eighteenth century, rent seeking had merely been redirected from local monopolies to state monopolies, and to Parliament.[144] And it was the promotion of institutional changes by the rent-seeking mercantilists, which encouraged the long-term expansion of the economy, as well as innovation.[145] And, it was believed that trade and industrial development was the cause of economic development, and not its consequence.[146] Another aspect of Mercantilism was that colonies were restricted to the supply of raw materials, while the mother country would export manufactured goods to colonies, and no manufacturing was allowed in colonies.[147] Moreover, the Mercantilists maintained that, if the country exported more than it imported, then there would be no serious drain on bullion reserves.[148] The increase in the exporting country's gold and silver reserves would cause the money supply to increase, increasing demand and supply, and thus increasing employment.[149] However, although the balance of trade principle was known long before Mun, it was he who linked the supply of money to the price level; and the detrimental effect of a high money supply, through higher prices, on the value of exports.[150] It was from this point of view that

[142] Heaton, H. (1937), Heckscher on Mercantilism, Journal of Political Economy, Vol. 45, No. 3, pp. 370–393.

[143] Johnson, E.A.J. (1928), Some Evidence of Mercantilism in the Massachusetts-Bay, The New England Quarterly, Vol. 1, No. 3, pp. 371–395.

[144] Mokyr, K., and Nye, J. (2007), Distributional Coalitions, the Industrial Revolution, the Origins of Economic Growth in Britain, Southern Economic Journal, Vol. 74, No. 1, pp. 50–70.

[145] Ibid.

[146] Smith, A. (1776), The Wealth of Nations, reproduced 2014, CreateSpace Independent Publishing Platform.

[147] Ibid.

[148] Ibid.

[149] Grampp, W. (1952), The Liberal Elements in English Mercantilism, The Quarterly Journal of Economics, Vol. 66, No. 4, pp. 465–501.

[150] Herlitz, L. (1964), The concept of mercantilism, Scandinavian Economic History Review, 12:2, 101–120, DOI: https://doi.org/10.1080/03585522.1964.10407639.

Mercantilists supported the export of bullion by merchants, by licence.[151] Mercantilists also recognised that the price of English cloth had to be competitive in order to mitigate the elasticity of foreign demand.[152] Nobility and statesman would see balance of trade and Mercantilist ideas as a source of prosperity to strength.[153] Foreign trade was then the cornerstone of Mercantilist ideology for two reasons[154]: first, it helped to maintain employment in England. The Mercantilists believed that free trade would reduce employment levels, as well as an inattention to the money supply.[155] Mercantile labour policy was based on the elements of economic psychology.[156] Employment would provide the stimulus of a physical environment, and it would drive peer competition for self-betterment and pecuniary gain.[157] Idleness, through unemployment should be avoided by getting the unemployed carry out meaningless manual tasks.[158] Secondly, foreign trade was a source of precious metals such as silver and gold, for a country which did not have many mines. Repeated economic crises, especially the one in 1620, weighed heavily in the favour of accumulating precious metals such as silver and gold by the state through tariffs on trade.[159] Before 1620, the main objective of trade policy was the prevention of exchange rate depreciation, whereas after 1620, this was no longer the case.[160] Mun hypothesised that the low value of England's exchanges after the economic crisis of 1620 was the result of the adverse balance of trade, and not is cause.[161] One

[151] Ibid.

[152] Ibid.

[153] Hinton, R.W.K. (1955), The Mercantile System in the Time of Thomas Mun, The Economic History Review, New Series, Vol. 7, No. 3, pp. 277–290.

[154] Ibid.

[155] Grampp, W. (1952), The Liberal Elements in English Mercantilism, The Quarterly Journal of Economics, Vol. 66, No. 4, pp. 465–501.

[156] Ibid.

[157] Ibid.

[158] Ibid.

[159] Hinton, R.W.K. (1955), The Mercantile System in the Time of Thomas Mun, The Economic History Review, New Series, Vol. 7, No. 3, pp. 277–290.

[160] Viner, J. (1930), English Theories of Foreign Trade Before Adam Smith, Journal of Political Economy, Vol. 38, No. 3, pp. 249–301.

[161] Gould, J.D. (1955), The Trade Crisis of the Early 1920's and English Economic Thought, The Journal of Economic History, Vol. 15, No. 2, pp. 121–133.

way in which to counter the negative effects of economic crises in sixteenth-century England was to plant colonies abroad, in the Americas and the West Indies, and boost trade with Africa, India, and other parts of Asia.[162] Wealth in the global economy was static, according to the Mercantilists, with some nations accumulating wealth at the expense of other nations.[163] Furthermore, the world was characterised by the scarcity of resources, as land was limited, and value and property were defined only with reference to land.[164] Vicious competition than characterised economic life.[165]

England's colonies in America also inherited the mercantilist tendencies of the mother country, which continued even after independence.[166] Moreover, the American revolutionaries considered their fledgling nation as an empire with a long-term view to westward expansion; and the eventual incorporation of Canada, West Indies, and Spanish Florida, and perhaps even Ireland.[167] Therefore, whether urban, rural, English, or American revolutionary, the mercantilists were nationalists in favour of national self-sufficiency, production of goods, and a positive balance of trade.[168] Moreover, the Mercantilists favoured not only a production economy, but also the control of export and raw material markets.[169] The basis of mercantilism was the protection and integration of domestic markets, while expanding abroad.[170] In this case, the mercantilists were not in favour of either a consumption dependent economy or on the economic interdependence of states, while a surplus of goods was viewed negatively.[171] The feudal fear of exports was side-lined by the mercantilists, in

---

[162] Hinton, R.W.K. (1955), The Mercantile System in the Time of Thomas Mun, The Economic History Review, New Series, Vol. 7, No. 3, pp. 277–290.

[163] Pincus, S. (2012), Rethinking Mercantilism: Political Economy, the British Empire, and the Atlantic World in the Seventeenth and Eighteenth Centuries, The William and Mary Quarterly, Vol. 69, No. 1, pp. 3–34.

[164] Ibid.

[165] Ibid.

[166] Williams, W. (1958), The Age of Mercantilism: An Interpretation of the American Political Economy, 1763 to 1828, The William and Mary Quarterly, Vol. 15, No. 4, pp. 419–437.

[167] Ibid.

[168] Ibid.

[169] Ibid.

[170] Ibid.

[171] Ibid.

favour of revenue generating taxes on imports.[172] However, Mercantilist trade barriers led to a loss of economic welfare in the traditional economic sense, due to the high prices paid by consumers, the misallocation of capital, as well as war.[173] Moreover, restrictions on imports into England had serious consequences. For example, in 1795 the effects of a disastrous wheat harvest could not be mitigated by importing wheat, which led to a rise in the price of bread, leading to social disorder.[174] At a religious level, the 'chosen people' fervour of the Puritans also fanned the flames of the Mercantilists' economic nationalism.[175]

The main feature of the Mercantilist system was the trading company, which was not only regulated by its members, but also by the Crown.[176] Laws were enacted that determined who could trade with whom, as well as who could transport British goods.[177] The merchant cartels of sixteenth-century England were essentially foreign trading companies, which were subject to extensive rules of governance to ensure that these merchant cartel companies were regulated companies.[178] However, after the mid-sixteenth century, public joint stock companies began to emerge. These companies were essentially no different from the regulated companies, except that the contractual innovation of the limited liability of investors made it easier for the public joint stock companies to raise financial capital.[179] The sustainability of increased foreign trade by the English merchant cartels required more and more financial resources to fund foreign expeditions. Ships provisioned with English goods and gold bullion had to be paid for, and the joint stock company was an effective way in which the financial capital required for this task could be raised. The financial

---

[172] Ibid.

[173] Pecquet, G.M. (2003), British Mercantilism and Crop Controls in the Tobacco Colonies: A Study of Rent-Seeking Costs, 22 Cato J. 467.

[174] Somers, M., and Block, F. (2005), From Poverty to Perversity: Ideas, Markets and Institutions over 200 Years of Welfare Debate, American Sociological review, Vol. 70, pp. 260–287.

[175] Williams, W. (1958), The Age of Mercantilism: An Interpretation of the American Political Economy, 1763 to 1828, The William and Mary Quarterly, Vol. 15, No. 4, pp. 419–437.

[176] Ibid.

[177] Levine, P. (2013), The British Empire – Sunrise to Sunset, 2nd Edition, Routledge, London.

[178] Ekelund, R., and Tollison, R. (1980), Mercantilist Origins of the Corporation, The Bell Journal of Economics, Vol. 11, No. 2, pp. 715–720.

[179] Ibid.

innovation of contractual limited liability allowed for increased international trade. Transferable property rights were also an innovation of the time. When cartel members died, there was a problem of what to do with their share of the business.[180] The early regulated companies specialising in trade with Africa exhibited this problem.[181] The problem was resolved by transferring cartel property rights of the member's share to family members.[182] However, it was not to be until the late nineteenth century that the limited liability of the joint stock companies and the property rights of cartel members were recognised by statute.[183]

References

Anderson, J. (1991), Is human cognition adaptive, Behavioural and Brain Sciences, 14, pp. 471–517.

Appleby, J. (1976), Ideology and Theory: The Tension Between Political and Economic Liberalism in Seventeenth Century England, The American Historical Review, Vol. 81, No. 3, pp. 499–515.

Audretsch, D., and Feldman, M. (2004), Knowledge Spillovers and the Geography of Innovation, IN Handbook of Regional and Urban Economics, Volume 4, Henderson, J., and Thisse, J. (Eds), Elsevier B.V.

Ayres, R.U. (1985), A Schumpeterian model of technological substitution. Technological Forecasting and Social Change 27: 375–383.

Bergek, A., Jacobson, S., Carlsson, B., Lindmark, S., and Rickne, A. (2008), Analysing the functional dynamics of technological innovation systems: A Scheme of Analysis, Research Policy, 37, pp. 407–429.

Blaug, M. (1964), Economic Theory and Economic History in Great Britain, 1650–1776, Past & Present, No. 28, pp. 111–116.

Burt, R. (1987), Social Contagion and Innovation: Cohesion versus Structural Equivalence,' American Journal of Sociology, 92, 1287–1335.

[180] Ibid.

[181] Scott. W.R. (1951), The Constitution and Finance of English, Scottish, and Irish Joint-Stock Companies to 1720, Vols. 1, 2. New York: Peter Smith.

[182] Ekelund, R., and Tollison, R. (1980), Mercantilist Origins of the Corporation, The Bell Journal of Economics, Vol. 11. No. 2, pp. 715–720.

[183] Ibid.

Cainelli, G., Evangelista, R., and Savona, M. (2006), Innovation and Economic Performance in Services: A Firm Level Analysis, Cambridge Journal of Economics, 30, pp. 435–458.

Carr, W. (2013), The Origins of the Wars of German Unification, Routledge, London.

Christensen, C. (2006), The Ongoing Process of Building a Theory of Disruption, J. Prod. Innov. Manag, 23, pp. 39–55.

Coleman, D. (1980), Mercantilism Revisited, The Historical Journal, Vol. 23, No. 4, pp. 773–791.

Coleman, J.S., Katz, E., and Menzel, H. (1966), Medical Innovation, Bobbs-Merrill, Indianapolis.

Coleman, J. S. (1986), Social theory, social research, and a theory of action, American Journal of Sociology, 91, 1309–1335.

Collier, J. (2006), Great Inventions–The Steam Engine, Marshall Cavendish Benchmark, New York.

Dooley, K., and Ven, A. (1999), Explaining Complex Organisational Dynamics, Organisational Science, Vol. 10, No. 3281–292.

Dosi, G. (1982), Technological Paradigms and Technological Trajectories: A Suggested Interpretation of the Determinants and Directions of Technological Change, Research Policy 11, pp. 147–162.

Ekelund, R., and Tollison, R. (1980), Mercantilist Origins of the Corporation, The Bell Journal of Economics, Vol. 11, No. 2, pp. 715–720.

Elmslie, B. (2015), Early English Mercantilists and the Support of Liberal Institutions, History of Political Economy, 47:3.

Ericsson, J., and Irandoust, M. (2001), On the Causality Between Foreign Direct Investment and Output: A Comparative Study, The International Trade Journal, Vol. XV, No. 1 1–26.

Etzkowitz, H., and Leydesdorff, L. (2000), The Dynamics of Innovation; From National Systems and 'Mode 2' to a Triple Helix of University-industry-government relation, Research Policy, 29, pp. 109–123.

Feldman, M.P. (1994a). The Geography of Innovation. Kluwer Academic, Boston.

Feldman, M. (1994b). "Knowledge Complementarity and Innovation". Small Business Economics 6 (3), 363–372.

Findlay, C., and Lumsden, C. (1988), The Creative Mind: Toward an Evolutionary Theory of Discovery and Innovation, J. Social Biol. Struct, 11, pp. 3–55.

Gaido, D. (2016), Rudolf Hilferding on English Mercantilism, History of Political Economy, 48:3.449–470.
Geroski, P., and Walters, C. (1995), Innovative Activity and the Business Cycle, The Economic Journal, Vol. 105, No. 431, pp. 916–928.
Gould, J.D. (1955), The Trade Crisis of the Early 1920's and English Economic Thought, The Journal of Economic History, Vol. 15, No. 2, pp. 121–133.
Grampp, W. (1952), The Liberal Elements in English Mercantilism, The Quarterly Journal of Economics, Vol. 66, No. 4, pp. 465–501.
Hage, J.T. (1999), Organisational Innovation and Organisational Change, Annual Reviews Sociology, Vol. 25, pp. 597–622.
Harkema, S. (2003), A Complex Adaptive Perspective on Learning Within Innovation Projects, The Learning Organization, Vol. 10 Issue: 6, pp. 340–346, doi:https://doi.org/10.1108/09696470310497177
Heaton, H. (1937), Heckscher on Mercantilism, Journal of Political Economy, Vol. 45, No. 3, pp. 370–393.
Hedstrom, P., Sandell, R., and Stern, C. (2000), Mesolevel Networks and the Diffusion of Social Movements: The Case of the Swedish Social Democratic Party, American Journal of Sociology, 106, pp. 145–172.
Hedstrom, P., and Wennberg, K. (2017), Causal Mechanism in Organisation and Innovation Studies, Innovation, 19:1, pp. 91–102.
Herlitz, L. (1964) The Concept of Mercantilism, Scandinavian Economic History Review, 12:2, 101–120, DOI:https://doi.org/10.1080/03585522.1964.10407639.
Hill, D. (1969), Richard Arkwright and the Water Frame, Proceedings of the Institution of Mechanical Engineers, Vol. 184, Issue 1 1175–1178.
Hills, R. (1970), Sir Richard Arkwright and his patent Granted in 1769, The Royal Society Journal of the History of Science, Vol. 24, Issue 2 254–260.
Hinton, R.W.K. (1955), The Mercantile System in the Time of Thomas Mun, The Economic History Review, New Series, Vol. 7, No. 3, pp. 277–290.
Hughes, T.P. (1983), Networks of Power: Electrification of Western Society 1880–1930, Johns Hopkins University Press, Baltimore, MD.
Hunt, S., and Morgan, R. (1996), The Resource-Advantage Theory of Competition: Dynamics, Path Dependencies, and Evolutionary Dimensions, Journal of Marketing, Vol. 60, No. 4, pp. 107–114.
Inoue-Nakamura, N., and Matsuzawa, T. (1997), Development of Stone Tool Use by Wild Chimpanzees, Journal of Comparative Psychology, Vol. III (2), pp. 159–173.
Irwin, D. (1996), Against the Tide: An Intellectual History of Free Trade. Princeton, NJ.: Princeton University Press.

Johnson, E.A.J. (1928), Some Evidence of Mercantilism in the Massachusetts-Bay, The New England Quarterly, Vol. 1, No. 3, pp. 371–395.
Kalat, J. (2007), Biological Psychology, Thomson Wadsworth, Belmont, California, USA.
Kingstone, A., Smilek, D., and Eastwood, J. (2008), Cognitive Ethology: A New Approach for Studying Human Cognition, British Journal of Psychology, 99, pp. 317–340.
Koob, A. (2009), The Root of Thought: Unlocking Glia – The Brain Cell That Will Help Us Sharpen Our Wits, Heal Injury and Treat Brain Disease, Pearson Education, New Jersey, USA.
Knight, K.D. (1963), A Study of Technological Innovation. The Evolution of Digital Computers, DPhil Thesis, Carnegie, Institute of Technology, Pittsburgh.
Levine, P. (2013), The British Empire – Sunrise to Sunset, 2nd Edition, Routledge, London.
Macleod, C. (1988), Inventing the Industrial Revolution, The English Patent System, 1660–1800, Cambridge University Press, Cambridge.
Malmberg, A., and Maskell, P. (2002), The Elusive Concept of Localisation Economies: Towards a Knowledge-Based Theory of Spatial Clustering, Environment and Planning, Vol. 34, pp. 429–449.
Markus, M., and Robey, D. (1988), Informational Technology and Organisational Change: Causal Structure in Theory and Research, Management Science, Vol. 34, No. 5, pp. 583–598.
McGuire, W.J. (1983a), Search for the Self: Going Beyond Self-esteem and the Reactive Self. In Zucker, R., Arnoff, J., and Rabin, A. (Eds), Personality and the Prediction of Behaviour, Academic Press, New York.
McGuire, W. (1983b), A Contextualist Theory of Knowledge: Its Implications For Innovation and Reform in Psychological Research, Advances in Experimental Social Psychology, Vol. 16.
Miettinen, R. (1999), The riddle of things: Activity theory and actor-network theory as approaches to studying innovations, Mind, Culture, and Activity, 6:3, 170–195, DOI:https://doi.org/10.1080/10749039909524725.
Mirza, R. (2007), The Rise and Fall of the American Empire: A Re-Interpretation of History, Economics and Philosophy:1492–2006, Trafford Publishing, Victoria, Canada.
Mokyr, K., and Nye, J. (2007), Distributional Coalitions, the Industrial Revolution, the Origins of Economic Growth in Britain, Southern Economic Journal, Vol. 74, No. 1, pp. 50–70.

Mone, M., McKinley, W., and Barker, V. (1998), Organisational Decline and Innovation: A Contingency Framework, The Academy of Management Review, Vol. 23, No. 1, pp. 115–132.
Monge, P., Cozzens, M., and Contractor, N. (1992), Communication and Motivational Predictors of the Dynamics of Organisational Innovation, Organisation Science, Vol. 3, No. 2 250–274.
Mumford, M. D., & Gustafson, S. B. (2007), Creative Thought: Cognition and Problem Solving in a Dynamic System. IN Creativity Research Handbook (Vol. 2, pp. 33–77), M. A. Runco (Ed.), Cresskill, NJ: Hampton.
Mumford, M., Hunter, S., and Byrne, C. (2009), What is the Fundamental? The Role of Cognition in Creativity and Innovation, Industrial and Organisational Psychology 2, pp. 353–356.
Mun, T., (1621) 1986, A Discourse of Trade, from England unto the East-Indies, New York: Augustus M. Kelley.
Murmann, J., and Frenken, K. (2006), Toward a Systematic Framework for Research on Dominant Designs, Technological Innovations, and Industrial Change, Research Policy, 35, pp. 925–952.
Nooteboom, B. (1994), Innovation and Diffusion in Small Firms: Theory & Evidence, Small Business Economics, 6, pp. 327–347.
Oers, B., Wardekker, W., Elbers, E., and Veer, R. (2008), The Transformation of Learning: Advances in Cultural-Historical Activity Theory, Cambridge University Press, Cambridge.
Pecquet, G.M. (2003), British Mercantilism and Crop Controls in the Tobacco Colonies: A Study of Rent-Seeking Costs, 22 Cato Journal 467.
Pincus, S. (2012), Rethinking Mercantilism: Political Economy, the British Empire, and the Atlantic World in the 17th and 18th Centuries, The William and Mary Quarterly, Vol. 69, No. 1, pp. 3–34.
Pradella, L. (2014), New Developments and the Origins of Methodological Nationalism, Competition and Change, Vol. 18, No. 2, pp. 180–193.
Rogoff, B. (2003), The Cultural Nature of Human Development, Oxford University Press, Oxford.
Room, G., Spence, A., Evangelou, E., Zeppini, P., Pugliese, E., Napolitano, L. (2017), Dynamics of Cumulative Innovation in Complex Social Systems (DCICSS). UK: University of Bath.
Ruef, M. (2002), Strong ties, weak ties and islands: Structural and cultural predictors of organisational innovation, Industrial and Corporate Change, Vol. 11, No. 3, pp. 427–449.
Russell, B. (2014), James Watt: Making the World Anew, Reaktion Books Ltd., London.

Saad, M. (2000), Development Through Technology Transfer: Creating New Organisational and Cultural Understanding, Intellect Books, Portland, Oregon.

Saltz, E. (1971), The Cognition Bases of Human Learning, Homewood, Dorsey Press.

Scherer, F.M. (1965), Invention and Innovation in the Watt-Boulton Steam-Engine Venture, Technology & Culture, Vol. 6, No. 2, pp. 165–187.

Schmoller, G. (1884), The Mercantile System and Its Historical Significance.

Schmookler, J. (1966), Invention and Economic Growth, Harvard University Press, Cambridge, MA.

Schumpeter, J.A. (2000), Entrepreneurship as Innovation. In Swedberg, R. (Ed), Entrepreneurship, Oxford University Press, Oxford pp. 51–75.

Scott, W.R. (1951), The Constitution and Finance of English, Scottish, and Irish Joint-Stock Companies to 1720, Vols. 1, 2. New York: Peter Smith.

Smith, A. (1776), The Wealth of Nations, Reproduced 2014, CreateSpace Independent Publishing Platform.

Somers, M., and Block, F. (2005), From Poverty to Perversity: Ideas, Markets and Institutions over 200 Years of Welfare Debate, American Sociological review, Vol. 70, pp. 260–287.

Strang, D., and Meyer, J. (1993), Institutional Conditions for Diffusion, Theory and Society, 22, pp. 487–511.

Van de Ven, A., and Huber, G. (1990), Longitudinal Field Research Methods for Studying Processes of Organisational Change, Organisation Science, Vol. 1, No. 3, pp. 213–219.

Van Kleeck, M. H., & Kosslyn, S. M. (1991), Visual Information Processing: A Perspective, IN: Attention and Performance XIV, Meyer, D., & Kornblum, S. (Eds.), Bradford Books/MIT Press, Cambridge, MA.

Viner, J. (1930), English Theories of Foreign Trade Before Adam Smith, Journal of Political Economy, Vol. 38, No. 3, pp. 249–301.

Williams, W. (1958), The Age of Mercantilism: An Interpretation of the American Political Economy, 1763 to 1828, The William and Mary Quarterly, Vol. 15, No. 4, pp. 419–437.

Wolfe, R. (1994), Organisational Innovation: Review, Critique and Suggested Research Directions, Journal of Management Studies, 31:3.

Woodman, R., Sawyer, J., and Griffin, R. (1993), Toward a Theory of Organisational Creativity, The Academy of Management Review, Vol. 18, No. 2, pp. 293–321.

3

Pre-History: Emergence and Palaeolithic to Bronze Age—10,000 BC to 800 BC

Shrew-like creatures that lived around 150 million years ago are the ancestors of humans and mammals.[1] According to the latter, aquatic worms that lived over 600 million years ago are the ancestors of all amphibians, mammals, birds, reptiles, and fishes. Furthermore, the bacteria-like organisms that lived over 3 billion years ago are the 'parents' of all living plants and animals.[2] The theory of evolution suggests that creatures evolve and acquire anatomical features and behaviours over time in response to their environment, as a means of adaptation, through processes associated with the biological pathways of genetic change. Due to the process of evolutionary change, after several successive splits from preceding family groups, the family 'Hominoidea' resulted. This family split over time to form the family groups of 'Hominidae' and 'Pongidae'. These families sub-divided into 'Homo' and 'Pongo, Gorilla, and Pan' respectively. Some studies have found that the divergence time between

[1] Ayala, F., and Cela-Conde, C. (2017), Processes in Human Evolution: The journey from early hominins to Neanderthals and modern humans, Oxford University Press, New York.

[2] Ayala, F., and Cela-Conde, C. (2017), Processes in Human Evolution: The journey from early hominins to Neanderthals and modern humans, Oxford University Press, New York.

S. Ramesh, *The Rise of Empires*, https://doi.org/10.1007/978-3-030-01608-1_3

'Homo' and the great apes is between 11 and 14 million years.[3] On the other hand, other studies have found that the divergence time between 'Homo' and 'Gorilla' to be about 7.2 million years, while the divergence time between 'Homo' and 'Pan' (chimpanzee) is about 4.7 million years.[4] The genome DNA of humans and chimpanzees is 98–99% similar.[5] According to the latter, the implication of this in difference is that since the divergence between the chimpanzees and humans, 3%—or 90 million nucleotides—has been replaced. The change in DNA nucleotides between chimpanzees and humans is mainly associated with non-coding DNA, which is associated with the development process.[6] The non-coding DNA which humans have, and which chimpanzees do not have, are associated with brain development.[7] Fossils found in Africa are attributed to Homo Habilis, reputed to the first incarnation of the genus 'Homo'. Homo Habilis had a bigger brain than another earlier hominin, the Australopthecines. Archaeological evidence suggests that the Australopthecines may have practiced butchery, as cut marks have been found on the fossil bones of animals.[8] The eating of meat by Australopthecines may have helped the evolutionary transition to Homo Habilis.[9] Although Australopthecines may have used crudely shaped rocks as knives to butcher animal carcasses—either hunted or the remains of kills by other animals—it is recognised that Homo Habilis was the first maker of stone tools.[10] However, there has been controversy surrounding

[3] Venn, O., Turner, I., Mathieson, I., de Groot, N., Bontrop, R., and McVean, G. (2014), Strong male bias drives germline mutation in chimpanzees, Science, 344 (6189): 1272–1275.

[4] Takahata, N., Satta, Y., and Klein, J. (1995), Divergence Time and Population Size in the Lineage Leading to Modern Humans, Theoretical Population Biology, Vol. 48, No. 2, pp. 198–221.

[5] Ayala, F., and Cela-Conde, C. (2017), Processes in Human Evolution: The journey from early hominins to Neanderthals and modern humans, Oxford University Press, New York.

[6] Ibid.

[7] Khaitovich, P., Hellman, I., Enard, W., Nowick, K., Leinweber, M., Franz, H., Weiss, G., Lachmann, M., and Paabo, S. (2005), Parallel Patterns of Evolution in the Genomes and Transcriptomes of Humans and Chimpanzees', Vol. 309, Science.

[8] Fleagle, J. (2013), Primate Adaptation & Evolution, Elsevier, London.

[9] Wrangham, R. (2010), Catching Fire: How Cooking Made Us Human, Profile Books Ltd, London.

[10] Bunney, S. (1988), Will the real Homo Habilis stand up? New Scientist, No.1636.

Homo Habilis, which may just not represent one hominin species, but several co-existing in Africa between 2 and 1.6 million years ago.[11] The latter suggests that due to this controversy, the theory that two-different species of early hominin gave rise to the next incarnations of the genus 'Homo', Homo Erectus and Homo Sapiens. However, supposedly distinct species, based on skeletal differences, such as Homo Rudolfensis, could easily be included within the classification of Homo Habilis if sexual dimorphism is accepted.[12] Sexual dimorphism refers to the physical differences between males and females in the same species. The next incarnation of the species in the genus 'Homo' was Homo Erectus. This species was regarded as the intermediate link between early hominins and Homo Sapiens.[13] However, according to the latter, Homo Erectus may be representative of itself and another species, Homo Ergaster. With the latter being the immediate predecessor to Homo Erectus, a species with new intellectual abilities and the capacity to leave Africa—the first hominin to do so—arose in Africa about 2 million years ago.[14] The latter suggests that the move out to Africa, first to Asia and then to Europe, occurred about 1 million years ago. Homo Erectus had a larger body frame compared to Homo Habilis, as well as a larger brain. The move out of Africa could have occurred because of the need for Homo Erectus to find new sources of food, perhaps due to a population explosion. In addition to extending its range beyond Africa and in comparison, to its predecessors, Homo Erectus exhibited innovative behaviour in many ways. These included the first use of fire, systematic hunting, systematic tool making, the use of a consistent location to live, and the first indication of an extended childhood.[15] The use of fire may have helped Homo Erectus cook meat. This allowed the meat to become more digestible, allowing nutrients in the meat to be absorbed and processed more easily by the

[11] Ibid.

[12] Arsuaga, J., and Martinez, I. (1998), The Chosen Species: The Long March of Human Evolution, Blackwell Publishing, Oxford.

[13] Lewin, R. (2005), Human Evolution: An Illustrated Introduction, Blackwell Publishing, Oxford.

[14] Ibid.

[15] Lewin, R. (2005), Human Evolution: An Illustrated Introduction, Blackwell Publishing, Oxford.

body.[16] This in turn facilitated the increase in the brain size of Homo Erectus. This gave this species increased cognitive capacity and, therefore, the ability to live a more complex life than its ancestors did. Homo Erectus was the first hominin species to exhibit a nomadic, hunter-gatherer existence. This kind of behaviour would diffuse across time to be exhibited by his direct descendant, Homo Sapiens. Homo Erectus was followed by Homo Heidelbergensis which was a species very similar skeletally to Homo Neanderthalensis.[17] This is perhaps because Homo Neanderthalensis is the direct descendant of Homo Heidelbergensis, and this species may have given rise to a third archaic human species, the Denisovans,[18] although there is not much fossilised evidence of the Denisovans except a finger bone from which DNA was extracted for analysis. Modern-day Southeast Asians share 5% of Denisovan DNA.[19] However, of the two species, Homo Neanderthalensis evolved in Europe over several hundred thousand years. The species managed to survive until up to 30,000–24,000 years ago, when the last of the species died in Iberia—to where the species had been pushed by the ever-expanding groups of Homo Sapiens, diffusing out of Africa through the near east, Anatolia, and then into Europe, Asia, and the rest of the world. On the other hand, Homo Neanderthalensis tended to stay in the valleys in which they were born and grew, although females may have moved from one group to another for mating practices.[20] The latter also suggests that Homo Neanderthalensis did not travel far in order to source raw materials, maintained a small social network, and may have had cannibalistic tendencies. The young of Homo Neanderthalensis matured more quickly than their Homo Sapien counterparts, typically by the age of 14.[21] As a

---

[16] Wrangham, R. (2010), Catching Fire: How Cooking Made Us Human, Profile Books Ltd., London.

[17] Sarmiento, E., Sawyer, G., and Milner, R. (2007), The Last Human: A Guide to Twenty-Two Species of Extinct Humans, Yale University Press, London.

[18] Coolidge, F., and Wynn, T. (2018), The Rise of Homo Sapiens: The Evolution of Modern Thinking, Oxford University Press, New York.

[19] Ibid.

[20] Ibid.

[21] Coolidge, F., and Wynn, T. (2018), The Rise of Homo Sapiens: The Evolution of Modern Thinking, Oxford University Press, New York.

group, according to the latter, Homo Neanderthalensis were less able to benefit from a division of labour, as the entire group would join in the hunt for animals for meat. However, there tended to be more of a division of labour amongst Homo Sapien groups. This ensured that, for Homo Sapiens, a diversity of food resources ranging from plant matter to meat could be accumulated over a period of time. The Homo Sapien population would then be better able to balance and change their dietary intake as dictated by the environment.[22]

The main anatomical differences between Homo Neanderthalensis and Homo Sapiens lies in the skeletal frame and the structure of the brain, although the brain of the former was bigger than that of the latter.[23] Homo Neanderthalensis was more stocky, shorter, and more muscular than Homo Sapiens. However, there are many differences between the structures of the brains of the two species. Compared to the brain of Homo Sapiens, the brain of Homo Neanderthalensis had smaller olfactory bulbs, larger occipital cortices, narrower orbitofrontal cortices, narrower temporal lobes with less forward-projecting temporal lobes.[24] The olfactory bulbs encoded smell information from the nose to be sent to the brain. The occipital cortices were involved in the processing of information by the eyes. The orbitofrontal cortices were involved in decision making. The cerebellum was involved in the coherent co-ordinating of limb movements. And the temporal lobes were involved in the processing and interpretation of sensory input so that it may be retained as memory. The structural differences between the brains of Homo Neanderthalensis and Homo Sapiens suggests that the former had a lower capacity for the retention in memory of sensory inputs, was less able to co-ordinate limb movements (was awkward), was indecisive, better able to process visual information but less able to smell 'danger'. In the context of differences in the structure of the brains of the two species, in a conflict between the two species, Homo Sapiens would have the advantage. This perhaps

[22] Ibid.

[23] Ibid.

[24] Wynn, T., Overmann, K., and Coolidge, F. (2016), The False Dichotomy; A Refutation of the Neanderthal Indistinguishability Claim, Journal of Anthropological Sciences, Vol. 94, pp. 201–221.

explains why Homo Neanderthalensis was unable to survive in the face of competition from Homo Sapiens. However, Homo Neanderthalensis did exhibit some recognisable innovation in the context of developing 'hafting'—attaching handles to tools.[25] The latter suggests that Homo Neanderthalensis also used the Levallois Technology, a method of using rocks to chip away at other rocks of a specific kind in order to create sharp edges for cutting.

While Homo Neanderthalensis and Homo Sapiens were co-existing contemporaries, Homo Erectus was the direct descendant of both species. This may have arisen because of location specific speciation in a group or multi-locational speciation. For Homo Neanderthalensis, speciation is more likely to have occurred in Europe from Homo Erectus, with Homo Heidelbergensis acting as an intermediary. On the other hand, for Homo Sapiens, it is likely that speciation occurred in a group in Africa from where Homo Sapiens moved out of Africa to inhabit other parts of the world. Location-specific speciation and multi-local specification represents two models of evolution.[26] On the balance, recent genetic evidence suggests that all the Homo Sapiens on the planet today evolved from a small group in Africa from between 180,000 years to 150,000 years ago.[27] This small group of early Homo Sapiens itself originated from one 'Adam' and 'Eve', with the 'Eve' appearing 80,000 years before the 'Adam'.[28] Homo Sapiens began to leave Africa 100,000–60,000 years ago and spread to Europe and Asia, reaching the Americas only 15,000 years ago.[29] The migration out of Africa may have started due to climatic changes.[30] A possible route out of Africa could have been the present Red Sea where, in those times, the sea levels would have been lower, allowing

[25] Coolidge, F., and Wynn, T. (2018), The Rise of Homo Sapiens: The Evolution of Modern Thinking, Oxford University Press, New York.

[26] Lewin, R. (2005), Human Evolution: An Illustrated Introduction, Blackwell Publishing, Oxford.

[27] Cartwright, J. (2016), Evolution and Human Behaviour; Darwinian Perspectives on the Human Condition, Palgrave Macmillan, New York.

[28] Wells, S. (2017), The Journey of Man: A Genetic Odyssey, Princeton University Press, Princeton.

[29] Cartwright, J. (2016), Evolution and Human Behaviour; Darwinian Perspectives on the Human Condition, Palgrave Macmillan, New York.

[30] Tierney, J., de Menocal, P., and Zander, P. (2017), A Climatic Context for the out of Africa migration, Geology, 45, 11, pp. 1023–1026.

a migration path for the first Homo Sapiens to emerge from Africa. Waves and waves followed over subsequent millennia, using the coasts of the landmass to get to Australia 40,000 years ago. Innovativeness and adaptability allowed the new Homo Sapien settlers to settle into diverse habitats—from extreme heat to the extreme cold of Siberia. Crude sewing needles were developed to produce the clothing, from animal skins, needed to survive in extreme sub-zero temperatures. Crude structures, in the modern context, were built from timber and animal hides as domiciles. The racial diversity seen in modern-day humanity is a relatively recent phenomenon in the context of human evolution. Up to 15,000 years ago, the migrating Homo Sapiens were of one type of colour and facial features.[31] Racial differentiation has only happened within the last 15,000 years.

Early Civilisations

Gobekli Tepe: 10,000 BC

It has long been contested that settled farming came before religion. Furthermore, the reasons for why Homo Sapiens as a hunter-gatherer took to farming and animal husbandry at a single location have been a subject of debate. However, the Neolithic archaeological site at Gobekli Tepe may finally be able to settle this debate. Before the discovery of Gobekli Tepe, it had been thought that the Neolithic Revolution had been a moment of 'genius' in which the domestication of wild plants and animals occurred at a single site in the lands of present-day southern Iraq.[32] Perhaps this flash of genius could have been brought about by environmental changes that limited the quantity of wild plants that could be collected for consumption as well as reducing the availability of wild animals, which could have migrated to other areas.

[31] Roberts, A. (2009), The Incredible Human Journey, Bloomsbury Publishing Plc.

[32] Mann, C. (2011), The Birth of Religion, National Geographic Magazine, http://ngm.nationalgeographic.com/print/2011/06/gobekli-tepe/mann-text

Gobekli Tepe is 15 km north of the city of Sanliurfa in the south-east of modern-day Turkey. Archaeologists consider Gobekli Tepe to be the most ancient archaeological site to be found anywhere in the world.[33] However, there may be even older archaeological sites which have yet to be found. When Gobekli Tepe was under construction between 9600 BC and 9500 BC, a mini ice age was coming to an end in Europe.[34] According the latter, Gobekli Tepe pre-dates the pyramids of ancient Egypt by at least 7000 years. The site constituted megalithic blocks, each weighing 7–50 tons in weight, with intricate carvings of animals, arms, hands, and clothing.[35,36] The carvings of animals depicted insects, snakes, bulls, foxes, boars, and gazelles.[37] This animal symbolism might suggest that the rituals and rites celebrated at Gobekli Tepe were perhaps associated with nature. Some of the megalithic stones were over 18 ft. in height.[38] It is suggestive that, in the area surrounding the megalithic stones, a large variety of ancient flint tools such as knives, axes, and spearhead points were also found. Moreover, at the site, archaeologists have also found a miniature Sphinx, with similar characteristics to the giant Sphinx of ancient Egypt, which was built seven thousand years later.[39] Nevertheless, the hunter-gatherers who built Gobekli Tepe did so without any knowledge of writing, pottery, and metal.[40] However, the builders of Gobekli Tepe were innovative enough to develop the technology associated with

[33] Debertolis, P., Gulla, D., and Savolainen, H. (2017), Archaeoacoustic Analysis in Enclosure D at Gobekli Tepe in South Anatolia, Turkey, History and Archaeology, The 5th Human and Social Sciences at the Common Conference.

[34] Collins, A. (2014), Gobekli Tepe: Genesis of the Gods: The temple of the Watchers and the Discovery of Eden, Bear & Company, Rochester, Vermont, USA.

[35] Vinter, J.C. (2011), Ancient Earth Mysteries, AEM Publishing.

[36] Notroff, J., Dietrich, O., Dietrich, L., Tvetmarken, C., Kinzel, M., Schlindwein, J., Sanmez, D., and Clare, L. (2017), Mediterranean Archaeology and Archaeometry, Vol. 17, No. 2, pp. 57–74.

[37] Ibid.

[38] Mann, C. (2011), The Birth of Religion, National Geographic Magazine, http://ngm.nationalgeographic.com/print/2011/06/gobekli-tepe/mann-text

[39] Vinter, J.C. (2011), Ancient Earth Mysteries, AEM Publishing.

[40] Mann, C. (2011), The Birth of Religion, National Geographic Magazine, http://ngm.nationalgeographic.com/print/2011/06/gobekli-tepe/mann-text

stone masonry.[41] Nevertheless, linguistic expression in written form did not emerge until 3000 BC, up to 6000 years after the completion of Gobekli Tepe.[42] It is now regarded as a site used for worship, celebration, and feasting, and not as a permanent home.[43] However, the earliest evidence of the earliest cultivated grain has been found in the region surrounding Gobekli Tepe.[44]

Gobekli Tepe was constructed at a time when the Neolithic population was increasing causing increased pressures, perhaps greater conflicts, on the social fabric of Neolithic society. This led to the development of rituals.[45] Perhaps to bring about a sense of 'community' and maintain the cohesiveness of that society. However, the rituals could have been a cultural innovation symbolising Homo Sapiens' conquest of nature.[46] It is not clear what kind of rituals were conducted at Gobekli Tepe, although no burials have been found on the site.[47] However, where areas of the site have been filled in with soil, fragments of human bone have been found, including three partially preserved skulls. These shows cut marks, suggesting that when the victim had been killed, that the skull was defleshed using flint tools.[48] The latter also suggests that the skulls had probably been hung from cords, with at least one skull modified using red ochre.. But it is open to question as to whether ritual killings and/or cannibalism took place at the site.

---

[41] Scranton, L. (2014), China's Cosmological Prehistory: The Sophisticated Science Encoded in Civilisation's Earliest Symbols, Simon & Schuster.

[42] Ibid.

[43] Schmidt, K. (2010), Gobekli Tepe – the Stone Age Sanctuaries: New results of ongoing excavations with a special focus on sculptures and high reliefs, Docomenta Praehistorica, Ljubljana, Vol. 37, pp. 239–256.

[44] Scranton, L. (2014), China's Cosmological Prehistory: The Sophisticated Science Encoded in Civilisation's Earliest Symbols, Simon & Schuster.

[45] Simmons, A. (2007), The Neolithic Revolution in the Near East; Transforming the Human Landscape, The University of Arizona Press, Tucson.

[46] Mann, C. (2011), The Birth of Religion, National Geographic Magazine, http://ngm.nationalgeographic.com/print/2011/06/gobekli-tepe/mann-text

[47] Gresky, J., Haelm, J., and Clare, L. (2017), Modified human crania from Gobekli Tepe provide evidence for a new form of Neolithic skull cult, Science Advances, Vol. 3, No. 6.

[48] Ibid.

The construction of Gobekli Tepe must have involved a sizable number of people over an extended period, a process of building and rebuilding over centuries, transporting heavy stones over a distance. The building of the religious architecture at the site can be split into two phases. The T-shaped monoliths incorporate the first phase of building and has been dated to the 10th millennium BC.[49] On the other hand, according to the latter, the second phase incorporated the monumental round-oval buildings, the older layer has been dated to the 9th millennium BC. Now all people around the world were living a hunter-gatherer existence, except in the fertile crescent.[50] However, some argue that smaller-scale projects of a nature similar to Gobekli Tepe were occurring in other parts of the world, independently, albeit perhaps not at the same time.[51] The fertile crescent represents the present-day Middle East, incorporating present-day Iraq, Syria, Palestine, and Egypt. It was in this part of the world that permanent village settlements came into existence, although Gobekli Tepe was a site where communal rituals were conducted.[52] So far, no evidence has been found to suggest that it was a site of permanent human habitation.

The round-oval buildings at the site were accumulated over a period of 1200 years, during what is known as the pre-pottery period.[53] The stones could have been moved by rolling and pulling them over a series of blocks. The enduring mystery of Gobekli Tepe is the contradiction between the level of technology and innovation used to build at the site and the apparent lack of technological sophistication in contemporary society, for example, no pottery.[54] However, at an intuitive level, in a

[49] Debertolis, P., Gulla, D., and Savolainen, H. (2017), Archaeoacoustic Analysis in Enclosure D at Gobekli Tepe in South Anatolia, Turkey, History and Archaeology, The 5th Human and Social Sciences at the Common Conference.

[50] Ibid.

[51] Stanish, C. (2017), The Evolution of Human Co-operation: Ritual and Social Complexity in Stateless Societies, Cambridge University Press, Cambridge.

[52] Debertolis, P., Gulla, D., and Savolainen, H. (2017), Archaeoacoustic Analysis in Enclosure D at Gobekli Tepe in South Anatolia, Turkey, History and Archaeology, The 5th Human and Social Sciences at the Common Conference.

[53] Notroff, J., Dietrich, O., Dietrich, L., Tvetmarken, C., Kinzel, M., Schlindwein, J., Sanmez, D., and Clare, L. (2017), Mediterranean Archaeology and Archaeometry, Vol. 17, No. 2, pp. 57–74.

[54] Constable, N. (2009), World Atlas of Archaeology, Thalamus Publishing.

hunter-gatherer society, why would anyone need pots? Food could be eaten from where it was either picked or killed; and storage could have been in pits, perhaps lined with some protective substance. To sustain the builders, a large and consistent food supply would have been required, assuming there were no breaks in the time taken to construct the monument at Gobekli Tepe. There seems to have been significant feasts at the site, perhaps as a reward for the work undertaken there by its builders over a millennium.[55] The latter suggests that the groups in the region surrounding the site may have had a common culture with its own norms and values. This could have led to segmentation within the groups, such that a smaller number of people, the 'nobility', controlled a much larger 'workforce'.[56] These cultural norms and values would have established standards by which the members of the groups would behave, as well as behaving with each other. In this case, it would have made sense to grow crops as well as maintain animals for slaughter near the building site. This would be logical explanation as to why hunter-gatherers made the transition to a settled life, laying the roots of early western civilisation. The Neolithic Revolution was important because location-specific crop cultivation and animal husbandry, perhaps in conjunction with a smaller-scale hunter-gathering existence,[57] suggests that this may have involved some form of division of labour amongst the members of the settlement. Some gathering berries, some gathering firewood, some tending animals, some making sure that the communal fire did not go out, some collecting water and some looking after the crops. This allowed the individual to benefit from a sustained nutritional diet. In the foetal stage and in early life this would have allowed for an improvement in brain development.[58] An improved diet would have led to people having an increased and

[55] Dietrich O., Notroff J., Schmidt K. (2017), Feasting, Social Complexity, and the Emergence of the Early Neolithic of Upper Mesopotamia: A View from Gobekli Tepe. In: Chacon R., Mendoza R. (eds) Feast, Famine or Fighting?, Studies in Human Ecology and Adaptation, Vol. 8. Springer, Cham.

[56] Ibid.

[57] Simmons, A. (2007), The Neolithic Revolution in the Near East; Transforming the Human Landscape, The University of Arizona Press, Tucson.

[58] Cakmak, I., Graham, R., and Welch, R. (2009), Agricultural and Molecular Genetic Approaches to Improving Nutrition and Preventing Micronutrient Malnutrition Globally, in 'Impacts of

improved access to a range of foods. It follows that the consumption of a diverse range of foods would be micronutrient rich, and this would allow the body to absorb a range of micronutrients—vitamins and minerals. The assimilation of these micronutrients by the body would allow the brain's potential to be more utilised. Learning would become more efficient, new ideas would arise, which could be exchanged and discussed with individuals in a growing population. In other words, there would be greater technological and social innovation as a result.[59] Thus, a combination of onsite crop cultivation, animal husbandry, the gathering of plants and fruits, the hunting of animals and fish would lead to increased cognitive capacity due to the development and more efficient use of brain tissue. This would have an immediate impact on innovation and learning and would not need long-term evolutionary development over millennia. Therefore, it would be justifiable that the roots of human civilisation lie in the development of the human mind, the ability to absorb facts related to the environment, form ideas and then project these ideas into reality by implementing practical applications.

Sumer: 3500 BC

Between 700,000 BC and 9000 BC, there were no permanent settlements of Homo Sapiens. Existence was at the hunter-gatherer level, in which groups would move from one location to another gathering plants and seeds and hunting wild animals. Furthermore, technology remained at a basic level and it involved the use and the manufacture of stone implements.[60] According to the latter, the earliest permanent settlements began in the near east in the Neolithic period at around 9000 BC. These settlements were based around the Tigris and the Euphrates rivers.[61] Subsequently, this form of settlement spread to the rest of the world,

---

Agriculture on Human Health and Nutrition – Volume 1', Cakmak, I., and Welch, R. (Eds), Eolss Publishers Co. Ltd., United Kingdom.

[59] Mann, C. (2011), The Birth of Religion, National Geographic Magazine, http://ngm.nationalgeographic.com/print/2011/06/gobekli-tepe/mann-text

[60] Nemet-Nejat, K. (1998), Daily Life in Ancient Mesopotamia, Greenwood Press, London.

[61] Hunter, N. (2015), Daily Life in Ancient Sumer, Heinemann-Raintree, Chicago.

including China, by 4000 BC. The most important feature of these earliest forms of permanent settlement was the onsite cultivation of crops and animal husbandry. In a contemporary sense, these early permanent settlements were made up of a few hundred individuals. However, as time went by, individual settlements began to engage in trading activities and, as a result, more complex forms of social organisation evolved.[62] The latter suggests that, in the period 6000 BC to 4000 BC, specific cultures emerged in northern Mesopotamia. These cultures included the Hassuna, Samarra, and the Halaf. In the period of the Hassuna culture, spindle whorls were used to produce cloth, gems evidenced trade, and private property rights existed.[63] This was denoted by the stamp of a seal as proof of ownership. Buildings were either connected with a central courtyard or were separate. Some of the larger buildings were designated for religious or ceremonial purposes.[64] The period of the Samarra culture was defined by its elaborately decorated pottery and architecture, with buildings displaying wall junctions and external buttresses.[65] According to the latter, the period of Halaf culture was characterised by workshops for pottery production, indicating the specialised division of labour, and villages had cobbled streets. It was in the Halaf culture that the tournette was invented, to be followed by the potter's wheel in the following Ubaid period between 4000 BC and 3500 BC.[66] The invention and introduction of the potter's wheel into pottery production increased production through the division of labour and allowed for the production of pottery of a uniform quality. The Ubaidian was the largest ethnic group of Mesopotamia, having settled there in small southern fishing villages that then expanded into sizable urban centres for the time, such as Eridu, Ur, and Uruk, which was the largest of the three. Other cities included Umma and Lagash; together with Eridu, Ur, and Uruk, these cities exercised considerable economic, social, and political control over their respective

[62] Nemet-Nejat, K. (1998), Daily Life in Ancient Mesopotamia, Greenwood Press, London.

[63] Ibid.

[64] Ibid.

[65] Ibid.

[66] Ibid.

surrounding areas. As such, they formed the city states of the ancient Sumerian civilisation.[67] The latter suggests that Sumerian cities were walled, with a temple at the centre of the city. The king was divine and Sumerian society was segmented into nobles, commoners, and slaves. The nobility included members of the royal family as well as the priests. The commoners included the people who worked for the palace and the temple, as well as tradespeople, crafts people, farmers, and fishermen.[68] The latter suggests that 90% of the population were engaged in farming. The invention and exploitation of irrigation allowed the Sumerians to maximise the harvest potential of the nutrient rich soil which composed their land.[69] The use of irrigation and the rich fertile land allowed the Sumerians to produce surpluses of wheat and barley. These surpluses could be used for trade or to distribute to people when the harvest failed. However, while the economy was primarily agricultural, the Sumerians did produce pottery, textiles, and metal work. The Sumerians also imported copper, tin, and timber in exchange for wheat, barley, metal goods, and dried fish.[70] They also traded with the peoples of the Mediterranean and with Meluhha—the Sumerian name for the Indus Valley civilisation, which would have been in modern-day northern India.[71] During the Harappan phase (2600 BC to 2500 BC), the people of the Indus Valley built ships with a 'house' on the deck and steered with an oar at the helm.[72] Using such ships, the people of the Indus Valley could traverse waterways and seas in order to trade with the Sumerian/Babylonian merchants and traders. The Sumerians valued the individual and the individual's achievement.[73] This ideal was enforced by laws,

[67] Duiker, W., and Spielvogel, J. (2007), World History, Thomson-Wadsworth.

[68] Ibid.

[69] Kramer, S. (1963), The Sumerians: Their History, Culture and Character, The University of Chicago Press, London.

[70] Ibid.

[71] Crawford, H. (1991), Sumer and the Sumerians, Cambridge University Press, Cambridge.

[72] Parpola, A. (2015), The Roots of Hinduism: The Early Aryans and the Indus Civilisation, Published to Oxford Scholarship Online: August 2015, DOI: https://doi.org/10.1093/acprof:oso/9780190226909.001.0001

[73] Ibid.

which the Sumerians were the first to enact anywhere in the world. The Sumerians were also able to express their language in written form. Possibly the first to do so in history. This was often on clay tablets, some of which have survived to this day. The combination of a consistent diet, the emphasis on individual achievement, and a written language allowed the Sumerians to be highly innovative. They are reputed to have invented the wheel and to have been the first to divide the day into hours and minutes.[74]

Tepe Gawra and Uruk were the dominant cultures of ancient Mesopotamia, with the former being prevalent in the north and the latter in the south of the country.[75] This latter suggests that, between 3000 BC and 2900 BC, the empty northern regions had been settled by Sumerian speakers. In the Uruk region, large-scale building projects were undertaken. This would suggest that there was a small elite in control of a larger supply of craftsmen and builders. The number of buildings in the Uruk region were greater than those in the Tepe Gawra region. It follows that the Uruk, Babylonian civilisation was more complex in organisation.[76] This can be attributed to its large population density and high level of productivity. By 3000 BC, the language spoken in Mesopotamia was Sumerian, with the dominant ethnic group being the Sumerians. Writing on clay tablets was invented in the period 3200 BC to 2900 BC, the Protoliterate period, (Nemet-Nejat 1998). It was initially used for accounting purposes. During the Protoliterate period, the main centres of economic, political, and social activity were the temples. The temples owned the land, hired labourers and specialist craftsmen, was responsible for co-ordinating trade, and exercised political control across the city and its surroundings.[77] The latter suggests that each city was a self-contained irrigated environment surrounded by dessert and swamps. These cities would eventually become city states which were loosely connected through trade.

---

[74] Hunter, N. (2015), Daily Life in Ancient Sumer, Heinemann-Raintree, Chicago.

[75] Nemet-Nejat, K. (1998), Daily Life in Ancient Mesopotamia, Greenwood Press, London.

[76] Ibid.

[77] Ibid.

Sargon the Great (2334 BC to 2279 BC) was the first king to unite the city states of Mesopotamia; and by establishing his capital in the city of Akkad, his empire was named the Akkadian Empire. Sargon was the first ruler of a unified state to have a standing army and military installations dotted across his lands. The city of Akkad was one of the wealthiest cities of the ancient world.[78] The Akkadian Empire was located to the north of the Sumerian city states; and it conquered these states in 2340 BC.[79] The Akkadian Empire was ended by the Gutians, invaders from Southeast Kurdistan.[80] This happened in 2100 BC, after which the Sumerian city states regained their autonomy and continued to co-exist while warring with each other.[81] However, according to the latter, the Sumerian city states were again unified by Hammurabi in 1792 BC. As king, Hammurabi built more irrigation channels, temples, and defensive walls.[82] Hammurabi also established a more robust codified system of law, now known as the Code of Hammurabi. These laws were received by Hammurabi from the Sun god in 2250 BC, and were found inscribed on a black diorite, 8 ft. high and about 6 ft. in circumference, in the ancient city of Susa in 1901.[83] The latter suggests that the Code of Hammurabi was written at a period of world history when the people of Western Europe were living far less sophisticated lives. The Code of Hammurabi was very comprehensive in covering the economic, social, and cultural lives of the citizens of Sumer. For example, the Code dealt with Family Law, the amounts of fees and salaries which should be paid to various occupations, compensation for merchants robbed by marauders, and the need for written contracts to facilitate transactions.[84] The Code of Hammurabi also specifies the social order of Babylonian society. This was structured into three levels.[85] The

[78] Ibid.

[79] Duiker, W., and Spielvogel, J. (2007), World History, Thomson-Wadsworth.

[80] Saggs, H. (2000), Peoples of the Past: Babylonians, University of California Press, Berkeley, Los Angeles.

[81] Duiker, W., and Spielvogel, J. (2007), World History, Thomson-Wadsworth.

[82] Ibid.

[83] McNeil, D. (1967), The Code of Hammurabi, 53 A.B.A. J. 444.

[84] Ibid.

[85] Bertman, S. (2003), Handbook to Life in Ancient Mesopotamia, Oxford University Press, Oxford.

first tier included members of land-owning families, the Awilum. The second tier included all freemen who did not own any land, the Mushkenum. The lowest, third tier included the slaves who owned neither their freedom nor land, the Wardum. According to the Code, inter-tier crime was less severely punished than was intra-tier crime. However, at the same time, harsher penalties were specified for the privileged in Babylonian society who infringed on any aspect of the Code, in contrast to the less privileged.[86] The Code of Hammurabi also increases the opportunity for social mobility, perhaps through marriage. Nevertheless, property ownership became more concentrated during the reign of Hammurabi, and thus, political power.[87]

Early Dynastic Period Egypt: 3100 BC

The ancient history of Egypt spans from 2950 BC to 30 BC. It encompasses thirty-one dynasties, not including the Macedonian Dynasty (332 BC to 309 BC) and the Ptolemaic Period (309 BC to 30 BC). The First Dynasty was established in 2950 BC with the unification of Egypt by Namer; and there are three other notable dynasties. Firstly, the Twelfth Dynasty (1938 BC to 1755 BC) included the reign of Amenemhat IV (1770 BC to 1760 BC), during which time the Shang Dynasty was founded in China in 1766 BC.[88] The Eighteenth Dynasty (1190 BC to 1069 BC) includes the reign of Ramesses IV (1156 BC to 1150 BC), which encompasses the collapse of the Mycenaen Civilisation in the Mediterranean.[89] The Nineteenth Dynasty encompassed the reign of Ramesses II (1279 BC to 1213 BC), also known as Ramesses the Great. The Macedonian Dynasty encompassed the time of Alexander the Great; and the Ptolemaic Period encompasses the time of Cleopatra (51 BC to 30 BC) as well as the Roman invasion of Britain, which at that time was

---

[86] Ibid.

[87] Ibid.

[88] Wilkinson, T. (2010), The Rise and Fall of Ancient Egypt: The History of a Civilisation from 3000 BC to Cleopatra, Bloomsbury, London.

[89] Ibid.

occupied by native British tribes such as the Iceni and the Trinovante. These tribes revolted against Roman rule in 60 AD, following the invasion of Britain by Julius Caesar in 54 AD.[90]

Due to its unique geographical position, Egypt has served as part of the migratory route for hominin species leaving East Africa to the other parts of the Old World.[91] However, according to the latter, no human presence has been detected in Egypt between 11,000 BC and 8000 BC. Nevertheless, the earliest Neolithic cultures emerged in the western desert.[92] The latter suggests that short-term camps were set up at Nabta Playa and Bir Kiseiba by hunter-gatherers in the early Neolithic Period, between 8800 BC and 6800 BC. At Bir Kiseiba there is also evidence of well digging, semi-domestication of cattle for milk and blood, and the use of primitive grinding equipment.[93] However, the people still led a hunter-gatherer existence despite the evidence of semi-permanent settlement. Moving from the Middle Neolithic (6500 BC to 5100 BC) to the Late Neolithic Period (5100 BC to 4700 BC), the human occupation of the western desert began to increase, perhaps because of an exceptionally wet period in its history. Moving through this period, there is also evidence of an increase in the number of wells as well as an increase in wattle and daub constructions.[94] The earliest evidence for crop cultivation and animal husbandry, perhaps in permanent settlements, is to be found in the Badarian Culture, named after a site found near Sohag, of Upper Egypt.[95] So, around 5000 years ago, people began to settle and build simple homes along the fertile land surrounding the River Nile, still the world's longest river.[96] The latter suggests that, for these early ancient

[90] Barker, H. (2004), Iceni, Bladud Books, Bath, United Kingdom.

[91] Hendrickx, S., and Vermeersch, P. (2000), Pre-history: From the Palaeolithic to the Badarian Culture (700,000 to 4000 BC, IN: The Oxford History of Ancient Egypt, Shaw, I. (Ed), Oxford University Press, New York.

[92] Ibid.

[93] Ibid.

[94] Hendrickx, S., and Vermeersch, P. (2000), Pre-history: From the Palaeolithic to the Badarian Culture (700,000 to 4000 BC, IN: The Oxford History of Ancient Egypt, Shaw, I. (Ed), Oxford University Press, New York.

[95] Ibid.

[96] Deady, K. (2012), Great Civilisations, Ancient Egypt, Beyond Pyramids, Capstone Press, Minnesota.

Egyptians, life changed from a hunter-gatherer like existence to that of a farmer. People caught fish in the Nile and still hunted, but at the same time engaged in animal husbandry—cattle, sheep and goats—as well as crop cultivation along the land surrounding the banks of the River Nile. They grew crops such as wheat and barley, as well as cultivating fruits and vegetables.[97] Over time, as more and more people began to settle, more and more villages and towns were established such that, between 3900 BC and 3100 BC, some of the towns became wealthy and powerful.[98] This second period of early Egyptian predynastic culture was known as the Naqada Period (4000 BC to 3200 BC), named after the site of Naqada in Upper Egypt.[99] The increase in population and the agglomeration of the area led to the formation of two distinct regions, Lower Egypt and Upper Egypt. The rulers of these regions fought each other for supremacy. However, in 3100 BC, King Narmer of Upper Egypt conquered Lower Egypt to form the unified state of Egypt.[100] The period encompassing the Egyptian Old Kingdom (2650 BC to 2134 BC) corresponded with the mature phase of urbanisation in the Indus Valley civilisation and the presence of the Akkadian Empire (2350 BC to 2100 BC). So, it is conceivable that trade took place between these three ancient civilisations. Indeed, there is some evidence for a 'village' of Indus Civilisation merchants in southern Iraq between 2114 BC and 2004 BC.[101]

In the early predynastic period, many innovations were introduced. The irrigation techniques were improved upon and, as a result, agricultural production increased.[102] During the period 3800 BC to 3500 BC, vessels made of stone and pottery were made in large numbers.[103]

---

[97] Ibid.

[98] Ibid.

[99] Midant-Reynes, B. (2000), Pre-history: From the Palaeolithic to the Badarian Culture (700,000 to 4000 BC: In The Oxford History of Ancient Egypt, Shaw, I. (Ed), Oxford University Press, New York.

[100] Deady, K. (2012), Great Civilisations, Ancient Egypt, Beyond Pyramids, Capstone Press, Minnesota.

[101] Lawler, A. (2008), Boring No More, a Trade-Savvy Indus Emerges, Science, Vol. 320, No. 5881, pp. 1276–1281.

[102] Shaw, I. (2012), Ancient Egyptian Technology & Innovation, Transformations in Pharaonic Material Culture, Bloomsbury, London.

[103] Ibid.

Furthermore, buildings made of mud brick were built during the early part of the Naqada period, 4000 BC to 3600 BC. But these buildings co-existed with buildings made of wattle and mud. Perhaps buildings made of wattle and mud were cheaper to produce than were buildings made of mud brick. Developments in metallurgy occurred by 4000 BC, with the smelting of copper to produce wood working tools.[104] The latter suggests that basket weaving techniques were applied to produce textiles using linen. The use of dyed thread and dyed cloth also began to make an appearance. Archaeological evidence also suggests that the ground or horizontal loom was used to weave textiles in the Badarian period between 4500 BC and 3800 BC. Knowledge of copper smelting technology was expanded to the use and production of other metals such as gold, silver, and lead, which were used to make artefacts, chariot parts, and weapons. It was during the early Badarian period that the methods of artificial mummification also began to emerge.[105] Many interrelated innovations in the lands of what would become dynastical ancient Egypt between 3400 BC and 3000 BC helped to establish a new society in around 3000 BC.[106] The first of these innovations was the invention of writing in 3250 BC.[107] However, this writing system has emerged as the ancient Egyptians had begun to use a pictorial system to denote meanings for their written language by 4000 BC.[108] There is some debate as to whether the ancient Egyptian writing system developed independently from the Sumerian writing system or if it developed as a result of the diffusion of knowledge. The latter is possible, given the geographical closeness of the two civilisations, which also overlapped in time. The southern Mesopotamian city of Uruk is most likely to have invented writing.[109]

---

[104] Ibid.

[105] Ibid.

[106] De Mieroop, M. (2011), A History of Ancient Egypt, John Wiley & Sons Ltd., Chichester, United Kingdom.

[107] Ibid.

[108] Shaw, I. (2012), Ancient Egyptian Technology & Innovation, Transformations in Pharaonic Material Culture, Bloomsbury, London.

[109] Wilkinson, T. (2010), The Rise and Fall of Ancient Egypt: The History of a Civilisation from 3000 BC to Cleopatra, Bloomsbury, London.

Over time, the idea, use, and benefits of writing diffused to the ancient Egyptians, who then set up their own pictorial system of writing. The knowledge of writing could have been spread from Uruk by travelling priests, scholars, merchants, or even through captured slaves and prisoners. As with all inventions throughout history, if the inventing society benefits from the discovery, other societies will adapt and use the invention for their benefit. The writing used in Uruk was in the form of simple pictorial depictions, dating to about 3100 BC.[110] The invention and use of writing can be ascribed to the ruler of Uruk Enmerkar, as stated in the ancient texts of Uruk. The invention of writing in Uruk can be attributed to the need to maintain a written record of the transactions through trade, as well to ascertain the levels of state inventories of grain, gold, holdings of land and slaves, and the wealth of the citizens. This would allow the state to be in a better position to extract taxes from the households of Uruk. The simple pictorial script of Uruk took a longer period to develop into Sumerian cuneiform.[111]

In ancient Egypt, writing developed further over the subsequent centuries until the world's first complete sentence was constructed in 2750 BC.[112] The benefits of writing are that it improves memory, allows the transmission of knowledge as well as its accumulation.[113] As a result, writing may help to indirectly stimulate the development of ideas based on existing knowledge. The oldest writing in ancient Egypt has been found in the predynastic tomb of King Abdju, an ancient Egyptian ruler who ruled 150 years before King Narmer unified Upper and Lower Egypt.[114] The writing found in the tomb was a short inscription, which was pictorial in nature, which evidenced the complexity that would be found in the hieroglyphics that would develop over the next 3500 years.[115]

---

[110] Dumper, M., and Stanley, B. (2007), Cities of the Middle East and North Africa: A Historical Encyclopaedia, ABC-CLIO Inc., Santa Barbara, California.

[111] Ibid.

[112] De Mieroop, M. (2011), A History of Ancient Egypt, John Wiley & Sons Ltd., Chichester, United Kingdom.

[113] Wilkinson, T. (2010), The Rise and Fall of Ancient Egypt: The History of a Civilisation from 3000 BC to Cleopatra, Bloomsbury, London.

[114] Ibid.

[115] Ibid.

The Egyptian form of writing was called hieroglyphics, and was written on papyrus—Egyptian paper—to keep records. However, the people of ancient Egypt also studied Mathematics, Science and Medicine and they built pyramids and obelisks.[116] These are still marvelled at in the modern world. Writing and papyrus also enabled the ancient Egyptians to make mathematical calculations—number operations, areas and angles.[117] The latter suggests that the ancient Egyptians may have used their knowledge of mathematics to aid their construction of the pyramids and obelisks. Along with limestone quarrying, pyramid construction began in the 3rd dynasty in the period 2686 BC to 2613 BC.[118]

Aegean Civilisation: 2700 BC

The civilisations of ancient Egypt, Mesopotamia, and the Indus Valley had arisen for centuries when Europe was still embracing the savagery of tribal hunter-gatherer society. It would be centuries before Europe would emerge from the darkness of barbarity into the twilight of civilisation on the shores of the Mediterranean.[119] The blossoming of culture in settlements, with people cultivating crops and engaging in animal husbandry—alongside the odd bit of hunting and gathering—began in the late Neolithic Period, at around 4000 BC.[120] At this time, the trajectory to the Mycenaean and Minoan civilisations was established.[121] The developing cultures of the islands of the Aegean and southern Laconia came to be at the forefront of development and innovation. The Minoan culture, associated with Knossos, would develop the potter's wheel for producing

[116] Deady, K. (2012), Great Civilisations, Ancient Egypt, Beyond Pyramids, Capstone Press, Minnesota.

[117] Shaw, I. (2012), Ancient Egyptian Technology & Innovation, Transformations in Pharaonic Material Culture, Bloomsbury, London.

[118] Ibid.

[119] Glotz, G. (1998), The Aegean Civilisation, Routledge.

[120] Van Andel, T., and Runnels, C. (1988), An essay on the 'emergence of civilisation' in Greece and the Aegean, Antiquity 62: 234–47.

[121] Ibid.

pottery.[122] This would signify increasing demand from a growing population being met with the increased specialisation and division of labour. This would require people to be trained, workshops to be built, and wheels produced. The Minoan civilisation would also produce distinct architecture on a big scale, record official transactions on documents, engage in long distance trade, engage in ritual practice, store agricultural surpluses centrally, and practice political organisation at a regional level.[123]

Following the rise of settlements and culture on the islands of the Aegean in 4000 BC, the settlements of central and north-eastern Greece began to fall behind, and would not be able to catch up for centuries.[124] Excavations at Dimini and Sesklo in Thessaly, mainland Greece, and Knossos on the island of Crete, have revealed stone tools and pottery made in the early Neolithic period.[125] Dimini was a settlement of 5000 BC. It covered a site with an area of one hectare, which was divided into concentric walls with a three-roomed building in a central court.[126] It has been thought that the central structure is for the elite.[127] The implication is that Dimini society was hierarchical.[128] The first known human occupation of Knossos occurred at around 7000 BC. The pottery of this age found on Knossos bears a distinctively decorated style that is found nowhere else in the Aegean. It must be concluded that Knossos culture developed in isolation from the rest of the Mediterranean.[129] Furthermore, archaeological evidence suggests that the people of Knossos kept domesticated animals

---

[122] Knappett, CJ. (1999), Tradition and Innovation in Pottery Forming Technology: Wheel-Throwing at Middle Minoan Knossos, BSA 94: 101–29.

[123] Ibid.

[124] Feuer, B. (1983), The northern Mycenaean border in Thessaly, British Archaeological Reports. International Series 176.

[125] Tomkins, P. (2010), Neolithic Antecedents, IN: The Oxford Handbook of the Bronze Age (ca. 3000–1000 BC), Cline, E.(Ed), Oxford University Press, New York.

[126] Tsountas, K. (1908). Ai Proistorikai Acropolis Diminiou Kai Sesklo, Athens: Athens, Archaeological Society.

[127] Halstead, P. (1993), Spondylus shell ornaments from late Neolithic Dimini, Greece: specialised manufacture or unequal accumulation? Antiquity 67, 603–9

[128] Fried, M.H. (1967), The evolution of political society, New York (NY): Random House.

[129] Isaakidou, V., and Tomkins, P. (2008), Escaping the Labyrinth: The Cretan Neolithic in Context, Oxbow Books.

and cultivated crops from the time of the earliest occupation of the site.[130] There has been controversy regarding the innovation of animal husbandry and crop cultivation creating agricultural economies, which over the centuries led to further innovation and development. The widely accepted hypothesis is that the practice of onsite animal husbandry and crop cultivation occurred in the near east first, and then spread to the rest of the world. However, there is no reason why crop cultivation and animal husbandry could not occur independently at different times in different places. This view is supported by some evidence. For example, the domestication of the dog occurred at different times in different parts of the world.[131] However, agricultural colonisation may have spread agricultural practice and innovations such as the use of the plough, as would have knowledge of the exploitation of secondary products such as milk and wool, which would have occurred in the period 3000 BC to 4000 BC.[132] As the population grew, land would have become more scarce, leading to a dispersal of habitation leading to greater inter-settlement trade rather than just intra-settlement trade.[133] Furthermore, the archaeological evidence of burial sites and stone axes does suggest that there was contact between the peoples of Knossos and those on mainland Greece.[134] The flint sickles have also been found at Dimini and Sesklo,[135] on the mainland, and at Demeter[136] on Knossos. At Dimini and Sesklo, archaeologists

[130] Ibid.

[131] Higgs, E., and Jarman, M. (1969), The Origins of Agriculture: A Reconsideration, Antiquity XLIII.

[132] Sherratt, A. G. (1981), Plough and pastoralism: aspects of the secondary products revolution. IN Pattern of the Past: Studies in Honour of David Clarke (eds I. Hodder, G. Isaac and N. Hammond). Cambridge: Cambridge University Press.

[133] Halstead, P. (1981), Counting sheep in Neolithic and bronze age Greece. IN Pattern of the Past: Studies in Honour of David Clarke (eds I. Hodder, G. Isaac and N. Hammond). Cambridge: Cambridge University Press, pp. 307–337.

[134] Higgs, E., and Jarman, M. (1969), The Origins of Agriculture: A Reconsideration, Antiquity XLIII.

[135] Runnels, C., and Murray, P. (2001), Greece Before History: An Archaeological Companion and Guide, Stanford University Press, Stanford, California.

[136] Hagg, R. (1998), Ancient Greek cult practice from the archaeological evidence: Proceedings of the Fourth International Seminar on Ancient Greek Cult, organized by the … Sueciae, series in 8) (Swedish Edition), Svenska Institutet i Athen.

have also found storage and cooking pots, querms for grinding corn, and knives made of obsidian.[137]

It was many centuries later, between 3000 BC and 2000 BC, that stone buildings and towns began to emerge, with the first palace on Knossos being built between 2000 BC and 1800 BC.[138] Minoan culture on Knossos was at its peak during the period 1600 BC to 1500 BC. Its architecture and decorative style were reflected across Crete and as far as Egypt.[139] By that period, two types of written archaic scripts had emerged. These included Cretan hieroglyphics and Linear A; both scripts of the Minoan/Knossos culture have not been deciphered yet.[140] Cretan hieroglyphics appeared for the last time at around 1500 BC. It was exclusively used in Crete, with its syllabary being composed of 90 symbols.[141] The first evidence of the use of Cretan hieroglyphics has been traced to the end of 3000 BC.[142] However, there is no evidence that suggest that it was used to convey a specific meaning.[143] Linear A was used more widely by people on the other islands of the Aegean, as well as in settlements on the mainland in southern Laconia, with its use continuing until 1450 BC.[144] It is believed that an archaic form of Linear A was used to inscribe the first text by stylus on clay tablets in 1800 BC, its syllabary being composed of 75 signs.[145] The third Aegean script was Linear B, derived from Linear A, which was used exclusively by the royalty for inscriptions on Mycenaean palaces—on Knossos in Crete as well as at other sites on the

[137] Runnels, C., and Murray, P. (2001), Greece Before History: An Archaeological Companion and Guide, Stanford University Press, Stanford, California.

[138] Scarre, C., and Stefoff, R. (2003), The Palace of Minos at Knossos, Oxford University Press, Oxford.

[139] Ibid.

[140] Ferrara, S. (2015), The beginnings of Writing on Crete: Theory and Context, The Annual of the British School at Athens, 110 (1), pp. 27–49.

[141] Oliver, J. (1986), Cretan Writing in the Second Millennium BC, World Archaeology, Vol. 17, No. 3, Early Writing Systems, pp. 377–389.

[142] Yule, P. (1980), Early Cretan Seals: A Study of Chronology, Philipp von Zabern Verlag, Mainz.

[143] Oliver, J. (1986), Cretan Writing in the Second Millennium BC, World Archaeology, Vol. 17, No. 3, Early Writing Systems, pp. 377–389.

[144] Ibid.

[145] Ibid.

mainland, such as Pylos and Mycenae.[146] Linear B was used until 1200 BC, and its syllabary was represented by 87 signs.[147] It is believed that Linear A was amended by the Mycenaeans in order to form Linear B, so that they could write their own language, Greek, in visual form.[148] A version of Cretan writing survived in Cyprus until 1000 BC, its syllabary being represented by 55 signs.[149] The development of a written form of language possibly started with a need to record, store, and exchange information. When records of transactions could be 'stored', even in symbolic form, taxes to be collected by the state from individuals and households could be more easily calculated. As the population of the Aegean expanded and settlements dispersed, inter-settlement trade expanded, generating information that needed to be recorded due to the limitations of human memory. Over time, more and more uses for the writing came into being, and it ceased to be used just for recording and for bronze stamps denoting either title or rank. Thus, there were economic reasons for the development of a written form of language over time.[150]

However, following an earthquake in 1480 BC, the Mycenaeans took control of the administration of Knossos from 1450 BC to 1380 BC, with control ending in 1250 BC, after which Knossos entered a phase of semi-habitation.[151]

Indus Valley Civilisation: 2600 BC

The Indus Valley civilisation was a contemporary civilisation with that of ancient Egypt and Mesopotamia (Sumer and Akkadian), and ancient China, (Dick 2006). However, while its contemporary civilisations

[146] Ibid.

[147] Ibid.

[148] Ibid.

[149] Ibid.

[150] Oliver, J. (1986), Cretan Writing in the Second Millennium BC, World Archaeology, Vol. 17, No. 3, Early Writing Systems, pp. 377–389.

[151] Scarre, C., and Stefoff, R. (2003), The Palace of Minos at Knossos, Oxford University Press, Oxford.

continued to prosper, the Indus Valley civilisation went into decline and eventually disappeared, leaving abandoned cities, towns, and villages, (McIntosh 2008). Nevertheless, based on the number of other excavated sites in the Indus Valley and its surrounding regions, the area covered by the Indus Valley civilisation extended 1100 km north to south; and 1600 km from west to east.[152] Some archaeologists contend that the Indus Valley civilisation may have been much bigger, extending as far as Mesopotamia.[153] Thus, at its height, the Indus Valley civilisation would have covered an area much bigger than the areas covered by either the civilisation of ancient Egypt or that of Mesopotamia (Sumer and Akkadian).

About 9000 years ago, people lived in the hills to the west of the Indus Valley, through which the River Indus flows.[154] The people were still hunter-gatherers, but kept animals such as cows for milk, However, sometime around 4000 BC, people began to cultivate crops on the nutrient rich lands surrounding and including the flood plains of the Indus River.[155] As a result, over the next few centuries, permanent settlements were established. By 3000 BC, a process of urbanisation had started. This can be associated with the establishment of cities such as Harappa and Mohenjo Daro. The former was situated on the banks of the River Ravi, the latter on land between the River Indus and the River Ghaggar-Hakra.[156] These ancient cities were 400 miles apart, but each had an identical architecture for the buildings, a layout of buildings and roads, with the same building technology having been used at both sites, (Sen 1988). However, Mohenjo Daro was slightly bigger than Harappa, and the city occupied 200 square hectares with a population of 20,000–40,000 people.[157] It would have been comparable to today's cities of Ur in Iraq and

[152] Sen, S. (1988), Ancient Indian History and Civilisation, New age International, New Delhi.

[153] Ibid.

[154] Dick, R. (2006), The Indus Valley Civilisation, Evan Brothers Limited, London.

[155] Ibid.

[156] Sen, B. (2011), Smart Green Civilisations: Indus Valley: Green Lessons from The Past, The Energy and Resources Institute, New Delhi.

[157] Lawler, A. (2008), Boring no more, a Trade Savvy Indus Emerges, Vol. 320, Science.

Memphis in Egypt.[158] Other sites associated with the Indus Valley civilisation include Amri and Chanhu-daro in southern Sind, Shortughai in north-east Afghanistan, and Lothal.[159] In Lothal, archaeologists found a brick dockyard that was linked to the Gulf of Cambay by a water channel.[160] In Rakhigarhi, another Indus city 340 km south-east of Harappa, archaeologists have found evidence of large-scale craft production.[161] Rakhigarhi was situated on the banks of the River Sarasvati. When this river dried up and disappeared, Rakhigarhi lost the source of its water and the people left the city, which decayed.[162] Archaeological evidence from Dholavira, another Indus city 800 km south of Harappa, suggests that the Indus Valley civilisation was based on a hierarchy of wealth rather than on totalitarianism.[163] This is evidenced by the fact that some homes had private wells and covered drains, while other homes had open drains and were next to cesspools.[164] It has also become evident that the people of Dholavira innovated and devised several hydraulic tools and strategies to ensure their city and land had a sufficient source of water because their city was not close to river as were some of the other cities of the Indus Valley civilisation.[165]

An analysis of the cities of Harappa and Mohenjo Daro illustrates the detailed and thorough town planning of the peoples of the Indus Valley civilisation. Each city had a citadel to the west, with residential town to the east.[166] The houses were built of mud brick, and the brick junctures was sealed with bitumen to ensure water resistance. Moreover, each house had a central courtyard, a water well, as well as a bathroom. The waste point of each bathroom was connected to a central sewer system that ran

---

[158] Ibid.

[159] Sen, S. (1988), Ancient Indian History and Civilisation, New age International, New Delhi.

[160] Ibid.

[161] Lawler, A. (2008), Boring no more, a Trade Savvy Indus Emerges, Vol. 320, Science.

[162] Sankhyan, A., and Rao, V. (2007), Human Origins, Genome and People of India; Genomic, Palaeontological & Archaeological Evidences, Allied Publishers Private Limited, New Delhi.

[163] Lawler, A. (2008), Boring no more, a Trade Savvy Indus Emerges, Vol. 320, Science.

[164] Ibid.

[165] Possehl, G. (2002), The Indus Civilisation: A Contemporary Perspective, Rowman & Littlefield Publishers Inc., London.

[166] Ibid.

at street level, covered by stone slabs or mud bricks.[167] The residential areas was laid out in a grid pattern with blocks of houses, connected by narrow roads, with no windows facing the road.[168] However, there was one main road which was about 30 ft. wide.[169] The cities also had granaries to stock the surplus of agricultural crops. Mohenjo Daro also had a great bath, which was presumably used for communal bathing. However, it may also have been used to collect rainwater, functioning as a water reservoir. But while neither city had a palace, as did the cities of other ancient civilisations, each city was surrounded by a very tall and very thick mud brick wall. This probably reflected the external dangers faced by the civilisations of the time.

Not only was the Indus Valley civilisation much bigger in terms of land area and population than its contemporary civilisations in ancient Egypt and Mesopotamia, it was also more advanced in some areas of engineering and technology.[170] The latter suggests that the Indus Valley civilisation was also a major driver in the world's first era of globalisation, 5000 years ago. In this case, it could trade goods with ancient Egypt and Mesopotamia. At the same time, this helped spread ideas through the movement of people from one civilisation to another such that an Indus translator went 'native' in Mesopotamia.[171] The key characterisation of Indus Valley civilisation was standardisation, from cities, to weights and bricks.[172] Moreover, as already discussed, the houses of the Indus Valley people were more sophisticated than the houses of the peoples of other contemporary civilisations, as they had wells to draw drinking water; and bathrooms which fed into the city's sewer system. This kind of homely sophistication would not occur again until the period of the Roman Empire (27 BC to 395 AD).[173] However, in contrast to its standardisation, oven baked bricks, and sewers, the Indus Valley civilisation did not

[167] Ibid.

[168] Ibid.

[169] Ibid.

[170] Lawler, A. (2008), Boring no more, a Trade Savvy Indus Emerges, Vol. 320, Science.

[171] Ibid.

[172] Ibid.

[173] Ibid.

leave behind any evidence of its cultural sophistication. There was no evidence of libraries with thousands of written clay tables, no temples, no palaces, no monuments, and no identifiable rulers of the time.[174] Furthermore, the written language of the Indus people has not been deciphered, and only consists of pictorial symbols on seals. No extensive form of the written language has yet been found. One possibility is that, unlike the written language of either Mesopotamia or ancient Egypt, the written language of the Indus Valley civilisation was put down on some type of paper which decayed over time, leaving no trace other than dust. Moreover, the sudden appearance of the Indus Valley cities on virgin land, some 600 years after the emergence of the first cities in Mesopotamia in 3200 BC, has puzzled archaeologists.[175] Here, the possibility arises that the peoples of the Indus Valley civilisation were either settlers from Mesopotamia, or were from some other unknown civilisation which archaeology has not yet found. However, recent excavations have found that some of the Indus cities were not built from scratch. For example, Harappa sits on a site which has been continuously occupied since 3700 BC.[176] Furthermore, archaeological evidence suggests that the pottery style was unique to the Indus Valley civilisation.[177] As was the religious practice, which seems to have revolved around goddess worship, tree worship, and the sacred bull. This has some similarities with the type of Hinduism practiced today in the villages of Gujarat in today's India, very close to the region which encompassed the Indus Valley civilisation.[178] In addition to having unique pottery and religious practices, the people of the Indus Valley civilisation learned how to mine metals such as copper and bronze, as well as then use these metals to produce weapons; and they also used polished metal as mirrors.[179] They also produced fish

---

[174] Ibid.

[175] Ibid.

[176] Ibid.

[177] Ibid.

[178] Robinson, A. (2015), The Indus: Lost Civilisations, Reaktion Books, London.

[179] Sen, B. (2011), Smart Green Civilisations: Indus Valley: Green Lessons from The Past, The Energy and Resources Institute, New Delhi.

hooks and razors using mined metal.[180] The world's oldest glass has been found at Harappa.[181] Other Indus Valley civilisations included the circular saw, saws with cutting 'teeth', and tubular drills, all of which would have been used for different types of woodworking.[182]

Neolithic China

Geographically, China can be divided between north and south. The division is created naturally by the Qinling Mountains. Northern China is dry and cool; and the climate is suitable for cultivating different type of crops, such as millets, barley, and soybeans.[183] On the other hand, the climate in southern China is much wetter, and the region is blessed with more lakes and rivers than the north.[184] Unlike the north, the south of China was home to a variety of ethnic peoples, including the Han, the majority ethnic group in China today. However, many of these ethnic peoples were either absorbed or displaced because of Han migration in conjunction with overwhelming and superior military force.[185]

When fossilised Homo Erectus fossils were found in a cave at Zhoukoudian, it was originally thought that the Han evolved from Homo Erectus. However, mitochondrial DNA analysis has proved that Han Chinese people are descended, like all peoples around the world, from Homo Sapiens who left their East African homeland about 100,000–70,000 years ago. Between 8000 BC and 3000 BC, the climate in China was much wetter than it is today. In addition, the geographical structure of the land was also very much different. For example, Shandong was cut off from the rest of the mainland by a sea dried up over time, and

[180] Lawler, A. (2008), Boring no more, a Trade Savvy Indus Emerges, Vol. 320, Science.

[181] Ibid.

[182] Agrawal, D. (2007), The Indus Civilisation: An Interdisciplinary Perspective, Aryan Books International.

[183] Tanner, H. (2010), China – A History – Volume 1 – From Neolithic Cultures through the Great Qing Empire, Hackett Publishing Company Inc., Indiana, USA.

[184] Ibid.

[185] Ibid.

became filled in.[186] It was in this environment that the ancestors of the modern Chinese took up farming and established permanent settlements. A warmer climate, a scarcity of edible wild plants, and the reduced number of animals to hunt, in conjunction with increasing population densities, may have created a need for people to find other sources of sustenance.[187] The people then turned to crop cultivation and animal husbandry. This could only be done efficiently in permanent fixed settlements. This gave rise to the emergence of agricultural settlements in the north and in the south of China. In the north, archaeological evidence from Henan, Hebei, Shaanxi, and Gansu supports the view of a dominant Peiligang Culture.[188] On the other hand, in the south, archaeological evidence from the provinces of Jiangxi, Guangxi, and Guangdong supports the view of a dominant Cord marked pottery culture.[189] The Peiligang Culture seems to have been the root of Chinese civilisation. This is because they buried their dead with turtle shells, and other possessions, which had a symbol inscribed on them. This could have been the beginning of ancestor worship, which is still practised in China today.[190] Turtle shells would then be used 6000 years later by the Shang people to conduct complex rites of divination; and the Peiligang symbols on the turtle shells resembled symbols in the Shang writing system, from which contemporary Chinese writing is derived.[191] While the symbols may not be sufficient evidence of the emergence of writing in Neolithic China, it could be a feature of the use of sign language, which eventually led to the development of a written language.[192] Writing in China developed over a period of time. It was not until the late Shang period (1200 BC to

---

[186] Ibid.

[187] Ibid.

[188] Tanner, H. (2010), China – A History – Volume 1 – From Neolithic Cultures through the Great Qing Empire, Hackett Publishing Company Inc., Indiana, USA.

[189] Ibid.

[190] Perkins, D. (1999), Encyclopaedia of China: History and Culture, Fitzroy Dearborn Publishers, New York, USA.

[191] Tanner, H. (2010), China – A History – Volume 1 – From Neolithic Cultures through the Great Qing Empire, Hackett Publishing Company Inc., Indiana, USA.

[192] Li, X., G. Harbottle, J. Zhang, and C. Wang. (2003), The earliest writing? Sign use in the seventh millennium BC at Jiahu, Henan Province, China. *Antiquity* 77 (295): 31–44.

1045 BC) that the earliest bronze—on objects belonging to either the nobility or the political elite—inscriptions appear.[193] The inscriptions usually denoted the name of the owner and his/her family, and the inscriptions were usually found on weapons or containers.[194] The earliest literature in China emerged during the Zhou dynasty. The literature was usually written on either silk or strips of bamboo and included historical records such as the 'Rites of Zhou' and poetry such as the 'The Book of Songs'.[195]

The Peiligang Culture was not the only one in Neolithic China. It is more likely that, in addition to the two mentioned cultures, these co-existed with other Neolithic Chinese cultures at the same time. These other cultures included the Longshan, Yangshao, Dawenkou, Xinle, and the Hongshan in the north, and the Majiabang, Hemudu, Daxi, Qujialing, and the Dapenkeng in the south.[196] Therefore, the impact on contemporary Chinese culture and society over time is most likely multicultural.

Two of the best-known Neolithic Chinese cultures are the Yangshao and the Longshan. The latter is the predecessor of the former. The Yangshao culture was based around the middle reaches of the Yellow River and the valleys of the River Wei.[197] This encompassed the modern-day Chinese provinces of Gansu and Henan. The Yangshao Culture existed between 5000 BC and 3000 BC. The innovative features of the Yangshao culture was the production of advanced stoneware and the practice of advanced agriculture.[198] The Yangshao culture cultivated dry crops such as millet and broomcorn millet.[199] All the tools used in the Yangshao culture were made of stone. Stone hammers were used for

[193] Moore, O. (2000), Reading the Past: Chinese, University of California Press, Los Angeles.

[194] Ibid.

[195] Duiker, W., and Spielvogel, J. (2007), World History, Thomson-Wadsworth.

[196] Ibid. Tanner, H. (2010), China – A History – Volume 1 – From Neolithic Cultures through the Great Qing Empire, Hackett Publishing Company Inc., Indiana, USA.

[197] Perkins, D. (1999), Encyclopaedia of China: History and Culture, Fitzroy Dearborn Publishers, New York, USA.

[198] Yinke, D. (2010), Ancient Chinese Civilisations, Cambridge University Press, Cambridge.

[199] The Editorial Committee of Chinese Civilisation. (2007), China: Five Thousand Years of History and Civilisation, City University of Hong Kong.

construction purposes. Thin stone shovels were used to dig the earth—for weeding and planting—and thin sickles with legs were used for harvesting crops.[200] For fishing, the peoples of the Yangshao culture used lances and hooks. These could have been made from stone or bone, as these have been found at archaeological sites. Similarly, for hunting animals, the peoples of the Yangshao culture used arrows, with inverted spikes, made from bone and stone.[201] The inverted design of the arrow head was innovative, as it would be deadlier to animals than the normal arrowhead. The Yangshao people were also able to spin and weave cloth at a primitive level, using spinning wheels made of pottery.[202] They were also able to sew primitive cloth and animal skins using bone needles. However, the Yangshao culture also exhibited very well-developed pottery making skills.[203] They also exhibited spatial and environmental considerations by being able to draw geometric figures and patterns in the shape of animals. The earliest copper artefacts unearthed in the world were found in Anatolia, dating from 9000 years ago.[204] However, the unearthed evidence of Yangshao bronze artefacts were dated to 3000 years later than the copper artefacts found in Anatolia. On the other hand, the Longshan Culture existed between 3000 BC and 1900 BC, along the middle and the lower reaches of the Yellow River. This encompasses the locations of the contemporary Chinese provinces of Shandong, Henan, Shanxi, and Shaanxi.[205] The Longshan Culture served as the platform for the early Chinese states and the Xia (2100 BC–1600 BC), Shang (1600 BC to 1050 BC), and Zhou (1046 BC to 256 BC) dynasties.[206] Archaeological evidence supports the view that the Longshan Culture used bronze implements made from

[200] Ibid.

[201] Ibid.

[202] Ibid.

[203] Yinke, D. (2010), Ancient Chinese Civilisations, Cambridge University Press, Cambridge.

[204] Ibid.

[205] Ibid.

[206] Liu, Li. (2004), The Chinese Neolithic: Trajectories to Early States, Cambridge University Press, Cambridge, England.

a combination of tin and copper. Some take the view that, as time progressed, this changed the hierarchical structure of society from a matriarchal one to a patriarchal one.[207] In other words, the use of metal tools in agriculture may have favoured males and disadvantage females. As a result, there was a divergence between the crafts sector (copper/bronze metallurgy, pottery, weaving, silk manufacture, jade carving) to the agricultural sector of the primitive economy.[208] However, there is no doubt that the Longshan Culture exhibited technological advancement in comparison to the Yangshao Culture. For example, the carving of Jade and the production of silk would have required specific knowledge and a certain level of technical sophistication. The Longshan Culture also exhibited a complex social organisation as well as class stratification.[209] This could have been due to differences in agricultural production between different farmers. Those farmers who produced more surpluses could accumulate wealth much faster. The wealth would have given these farmers a higher social status than farmers who did not produce as much. Over time, the differences in wealth in Longshan Culture could have given rise to nobility and common folk. Complex social organisation and class stratification could have been associated with increasing population, which would have stimulated increasing urbanisation and the emergence of walled towns. The fortified walls of Longshan cities were built using a technique known as 'hangtu'. This technique involved ramming earth into a dense structure for building the walls of cities as well as platforms in buildings.[210] The Longshan Culture shares many similarities with the later Shang and Zhou dynasties. The use of Jade in rituals, oracle bones in divination, well-made wheels, and standardised hard-fired pottery.[211]

[207] Ibid.

[208] Ibid.

[209] Dematte, P. (1999), Longshan-Era Urbanism: The Role of Cities in Predynastic China, Asian Perspectives, Vol. 38, No. 2, pp. 119–153.

[210] Ibid.

[211] Ibid.

References

Agrawal, D. (2007), The Indus Civilisation: An Interdisciplinary Perspective, Aryan Books International.

Arsuaga, J., and Martinez, I. (1998), The Chosen Species: The Long March of Human Evolution, Blackwell Publishing, Oxford.

Ayala, F., and Cela-Conde, C. (2017), Processes in Human Evolution: The journey from early hominins to Neanderthals and modern humans, Oxford University Press, New York.

Barker, H. (2004), Iceni, Bladud Books, Bath, United Kingdom.

Bertman, S. (2003), Handbook to Life in Ancient Mesopotamia, Oxford University Press, Oxford.

Bunney, S. (1988), Will the real Homo Habilis stand up? New Scientist, No.1636.

Cakmak, I., Graham, R., and Welch, R. (2009), Agricultural and Molecular Genetic Approaches to Improving Nutrition and Preventing Micronutrient Malnutrition Globally, in 'Impacts of Agriculture on Human Health and Nutrition – Volume 1,' Cakmak, I., and Welch, R. (Eds), Eolss Publishers Co. Ltd, United Kingdom.

Cartwright, J. (2016), Evolution and Human Behaviour; Darwinian Perspectives on the Human Condition, Palgrave Macmillan, New York.

Collins, A. (2014), Gobekli Tepe: Genesis of the Gods: The temple of the Watchers and the Discovery of Eden, Bear & Company, Rochester, Vermont, USA.

Constable, N. (2009), World Atlas of Archaeology, Thalamus Publishing.

Coolidge, F., and Wynn, T. (2018), The Rise of Homo Sapiens: The Evolution of Modern Thinking, Oxford University Press, New York.

Crawford, H. (1991), Sumer and the Sumerians, Cambridge University Press, Cambridge.

De Mieroop, M. (2011), A History of Ancient Egypt, John Wiley & Sons Ltd, Chichester, United Kingdom.

Deady, K. (2012), Great Civilisations, Ancient Egypt, Beyond Pyramids, Capstone Press, Minnesota.

Debertolis, P., Gulla, D., and Savolainen, H. (2017), Archaeoacoustic Analysis in Enclosure D at Gobekli Tepe in South Anatolia, Turkey, History and Archaeology, The 5th Human and Social Sciences at the Common Conference.

Dick, R. (2006), The Indus Valley Civilisation, Evan Brothers Limited, London.

Dietrich O., Notroff J., Schmidt K. (2017), Feasting, Social Complexity, and the Emergence of the Early Neolithic of Upper Mesopotamia: A View from Gobekli Tepe. In: Chacon R., Mendoza R. (eds) Feast, Famine or Fighting?, Studies in Human Ecology and Adaptation, Vol 8. Springer, Cham.

Duiker, W., and Spielvogel, J. (2007), World History, Thomson-Wadsworth.

Dumper, M., and Stanley, B. (2007), Cities of the Middle East and North Africa: A Historical Encyclopaedia, ABC-CLIO Inc, Santa Barbara, California.

Ferrara, S. (2015), The beginnings of Writing on Crete: Theory and Context, The Annual of the British School at Athens, 110 (1), pp. 27–49.

Feuer, B. (1983), The northern Mycenaean border in Thessaly, British Archaeological Reports. International Series 176.

Fleagle, J. (2013), Primate Adaptation & Evolution, Elsevier, London.

Fried, M.H. (1967), The evolution of political society, New York (NY): Random House.

Glotz, G. (1998), The Aegean Civilisation, Routledge.

Gresky, J., Haelm, J., and Clare, L. (2017), Modified human crania from Gobekli Tepe provide evidence for a new form of Neolithic skull cult, Science Advances, Vol. 3, No.6.

Hagg, R. (1998), Ancient Greek cult practice from the archaeological evidence: Proceedings of the Fourth International Seminar on Ancient Greek Cult, organised by the (Acta Instituti Atheniensis Regni Sueciae, series 8) (Swedish Edition), Svenska Institutet i Athen.

Halstead, P. (1981), Counting sheep in Neolithic and bronze age Greece. IN Pattern of the Past: Studies in Honour of David Clarke (eds I. Hodder, G. Isaac and N. Hammond). Cambridge: Cambridge University Press, pp. 307–339.

Halstead, P. (1993), Spondylus shell ornaments from late Neolithic Dimini, Greece: specialised manufacture or unequal accumulation? Antiquity 67, 603–9.

Hendrickx, S., and Vermeersch, P. (2000), Pre-history: From the Palaeolithic to the Badarian Culture (700,000 to 4000 BC, IN: The Oxford History of Ancient Egypt, Shaw, I. (Ed), Oxford University Press, New York.

Higgs, E., and Jarman, M. (1969), The Origins of Agriculture: A Reconsideration, Antiquity XLIII.

Hunter, N. (2015), Daily Life in Ancient Sumer, Heinemann-Raintree, Chicago.

Isaakidou, V., and Tomkins, P. (2008), Escaping the Labyrinth: The Cretan Neolithic in Context, Oxbow Books.

Khaitovich, P., Hellman, I., Enard, W., Nowick, K., Leinweber, M., Franz, H., Weiss, G., Lachmann, M., and Paabo, S. (2005), 'Parallel Patterns of Evolution in the Genomes and Transcriptomes of Humans and Chimpanzees,' Vol. 309, Science.

Knappett, CJ. (1999), Tradition and Innovation in Pottery Forming Technology: Wheel-Throwing at Middle Minoan Knossos, BSA 94:101–29.

Kramer, S. (1963), The Sumerians: Their History, Culture and Character, The University of Chicago Press, London.

Lawler, A. (2008), Boring No More, a Trade-Savvy Indus Emerges, Science, Vol. 320, Issue 5881, pp. 1276–1281.

Lewin, R. (2005), Human Evolution: An Illustrated Introduction, Blackwell Publishing, Oxford.

Li, X., G. Harbottle, J. Zhang, and C. Wang. (2003), The earliest writing? Sign use in the seventh millennium BC at Jiahu, Henan Province, China. *Antiquity* 77 (295): 31–44.

Mann, C. (2011), The Birth of Religion, National Geographic Magazine, http://ngm.nationalgeographic.com/print/2011/06/gobekli-tepe/mann-text

McIntosh, J. (2008), The ancient Indus Valley: New Perspectives, ABC-CLIO, Santa-Barbara, California.

McNeil, D. (1967), The Code of Hammurabi, 53 A.B.A. J. 444.

Midant-Reynes, B. (2000), Pre-history: From the Palaeolithic to the Badarian Culture (700,000 to 4000 BC): In The Oxford History of Ancient Egypt, Shaw, I. (Ed), Oxford University Press, New York.

Moore, O. (2000), Reading the Past: Chinese, University of California Press, Los Angeles.

Nemet-Nejat, K. (1998), Daily Life in Ancient Mesopotamia, Greenwood Press, London.

Notroff, J., Dietrich, O., Dietrich, L., Tvetmarken, C., Kinzel, M., Schlindwein, J., Sanmez, D., and Clare, L. (2017), Mediterranean Archaeology and Archaeometry, Vol. 17, No. 2, pp. 57–74.

Oliver, J. (1986), Cretan Writing in the Second Millennium BC, World Archaeology, Vol. 17, No. 3, Early Writing Systems, pp. 377–389.

Parpola, A. (2015), The Roots of Hinduism: The Early Aryans and the Indus Civilisation, Published to Oxford Scholarship Online: August 2015, doi: https://doi.org/10.1093/acprof:oso/9780190226909.001.0001.

Perkins, D. (1999), Encyclopaedia of China: History and Culture, Fitzroy Dearborn Publishers, New York, USA

Possehl, G. (2002), The Indus Civilisation: A Contemporary Perspective, Rowman & Littlefield Publishers Inc, London.
Roberts, A. (2009), The Incredible Human Journey, Bloomsbury Publishing Plc.
Robinson, A. (2015), The Indus: Lost Civilisations, Reaktion Books, London.
Runnels, C., and Murray, P. (2001), Greece Before History: An Archaeological Companion and Guide, Stanford University Press, Stanford, California.
Saggs, H. (2000), Peoples of the Past: Babylonians, University of California Press, Berkeley, Los Angeles.
Sankhyan, A., and Rao, V. (2007), Human Origins, Genome and People of India; Genomic, Palaeontological & Archaeological Evidences, Allied Publishers Private Limited, New Delhi.
Sarmiento, E., Sawyer, G., and Milner, R. (2007), The Last Human: A Guide to Twenty-Two Species of Extinct Humans, Yale University Press, London.
Scarre, C., and Stefoff, R. (2003), The Palace of Minos at Knossos, Oxford University Press, Oxford.
Schmidt, K. (2010), Gobekli Tepe – the Stone Age Sanctuaries: New results of ongoing excavations with a special focus on sculptures and high reliefs, Docomenta Praehistorica, Ljubljana, Vol. 37, pp. 239–256.
Scranton, L. (2014), China's Cosmological Prehistory: The Sophisticated Science Encoded in Civilisation's Earliest Symbols, Simon & Schuster.
Sen, B. (2011), Smart Green Civilisations: Indus Valley: Green Lessons from The Past, The Energy and Resources Institute, New Delhi.
Sen, S. (1988), Ancient Indian History and Civilisation, New Age International, New Delhi.
Shaw, I. (2012), Ancient Egyptian Technology & Innovation, Transformations in Pharaonic Material Culture, Bloomsbury, London.
Sherratt, A. G. (1981), Plough and pastoralism: aspects of the secondary products revolution. IN Pattern of the Past: Studies in Honour of David Clarke (eds I. Hodder, G. Isaac and N. Hammond). Cambridge: Cambridge University Press.
Simmons, A. (2007), The Neolithic Revolution in the Near East; Transforming the Human Landscape, The University of Arizona Press, Tucson.
Stanish, C. (2017), The Evolution of Human Co-operation: Ritual and Social Complexity in Stateless Societies, Cambridge University Press, Cambridge.
Takahata, N., Satta, Y., and Klein, J. (1995), Divergence Time and Population Size in the Lineage Leading to Modern Humans, Theoretical Population Biology, Volume 48, Issue 2, pp. 198–221.

Tanner, H. (2010), China – A History–Volume 1 – From Neolithic Cultures through the Great Qing Empire, Hackett Publishing Company Inc, Indiana, USA.

The Editorial Committee of Chinese Civilisation. (2007), China: Five Thousand Years of History and Civilisation, City University of Hong Kong.

Tierney, J., de Menocal, P., and Zander, P. (2017), A Climatic Context for the out of Africa migration, Geology, 45, 11, pp. 1023–1026.

Tomkins, P. (2010), Neolithic Antecedents, IN: The Oxford Handbook of the Bronze Age (ca. 3000–1000 BC), Cline, E. (Ed), Oxford University Press, New York.

Tsountas, K. (1908), Ai Proistorikai Acropolis Diminiou Kai Sesklo, Athens: Athens, Archaeological Society.

Van Andel, T., and Runnels, C. (1988), An essay on the 'emergence of civilisation' in Greece and the Aegean, Antiquity 62: 234–47.

Venn, O., Turner, I., Mathieson, I., de Groot, N., Bontrop, R., and McVean, G. (2014), Strong male bias drives germline mutation in chimpanzees, Science, 344 (6189): 1272–1275.

Vinter, J.C. (2011), Ancient Earth Mysteries, AEM Publishing.

Wells, S. (2017), The Journey of Man: A Genetic Odyssey, Princeton University Press, Princeton.

Wilkinson, T. (2010), The Rise and Fall of Ancient Egypt: The History of a Civilisation from 3000 BC to Cleopatra, Bloomsbury, London.

Wrangham, R. (2010), Catching Fire: How Cooking Made Us Human, Profile Books Ltd, London.

Wynn, T., Overmann, K., and Coolidge, F. (2016), The False Dichotomy; A Refutation of the Neanderthal Indistinguishability Claim, Journal of Anthropological Sciences, Vol. 94, pp. 201–221.

Yinke, D. (2010), Ancient Chinese Civilisations, Cambridge University Press, Cambridge.

Yule, P. (1980), Early Cretan Seals: A Study of Chronology, Philipp von Zabern Verlag, Mainz.

4

The Babylonian Empire: 1900 BC to 539 BC

The history of Sumer and Akkad have been mostly derived from archaeological evidence, unlike the history of either Greece or Rome which have been accorded a written tradition.[1] There was no Herodotus or Thucydides in the case of either Sumer or Akkad as there was in the case of Greece, except for Berossus.[2] However, he wrote in the latter Greek-Babylonian period, in the third century BC. Knowledge of Mesopotamian dynasties and kings have been mostly derived from inscribed tablets uncovered during archaeological excavations of ancient sites. Throughout the rise of Mesopotamian civilisation was the increased use of writing,[3] for recording transactions, but also as a tool of social and political control through written laws and regulations. Furthermore, there was also continuous innovation in the context of different forms of economic and political organisation.[4] However, as ancient Mesopotamia transitioned from Sumer to Assyria and Babylonia, the latter two civilisations inherited the

[1] Charpin, D. (1995), The History of Ancient Mesopotamia: An Overview, IN Sasson, J. (Ed), Civilizations of the Ancient Near East, Vol. 2, Simon and Schuster, New York.

[2] Ibid.

[3] Ibid.

[4] Ibid.

S. Ramesh, *The Rise of Empires*, https://doi.org/10.1007/978-3-030-01608-1_4

institutions and ideas of the former, but in a different language, Akkadian.[5] Furthermore, other technological innovations such as the horse drawn chariot and iron metallurgy had already taken place by 1500 BC and 1000 BC respectively.[6]

Historians of ancient Mesopotamia have given thought to state formation of pre-state societies bordering contemporary civilisations, as well to the case of state collapse.[7] Such a state collapse engulfed the Akkadian empire circa 2200 BC.[8] Some of the archaeological evidence relating to the resolution of these questions was focused on the ancient site of Tell Leilan, which was one of the three urban centres of its time on the Haberl plains.[9] Tell Leilan was continuously occupied from around 5500 BC, and became the political centre of the region from 1500 BC to 1000 BC.[10]

The Amorites: 1900 BC

To the west of the Euphrates River were located the Amorites. When the Sumerian dynasty collapsed in 2000 BC, the region plunged into chaos and disorder.[11] This may have been the result of a harsh and long drought which caused the agricultural economy of the Sumerians to collapse.[12] Akkadian cities in the northern plains were abandoned, and there was a mass migration of peoples to the south.[13] This caused greater demand for

[5] Grayson, A. (1980), Assyria and Babylonia, Orientalia, NOVA SERIES, Vol. 49, No. 2, pp. 140–194.

[6] Charpin, D. (1995), The History of Ancient Mesopotamia: An Overview, IN Sasson, J. (Ed), Civilizations of the Ancient Near East, Vol. 2, Simon and Schuster, New York.

[7] Weiss, H., Courty, M., Wetterstrom, W., Guichard, F., Senior, L., Meadow, R., and Curnow, A. (1993), The Genesis and Collapse of the Third Millennium North Mesopotamian Civilisation, Science, New Series, Vol. 261, No. 5124, pp. 995–1004.

[8] Ibid.

[9] Ibid.

[10] Ibid.

[11] De Lafayette, M. (2014), Thesaurus Lexicon of Similar Words & Synonyms in 21 Dead and Ancient Languages, Volume II 'A' 9Alma-Azu), Times Square Press, New York.

[12] Wilford, J. (1993), Collapse of Earliest Known Empire Is Linked to Long, Harsh Drought, The New York Times.

[13] Ibid.

scarce food and water in the south, leading to the fall of the Sargon dynasty.[14] The world's first known empire.[15] It was at this time that the Amorites invaded Mesopotamia and, by taking advantage of the chaos, they became dominant.[16] The Amorites were pastoralists, and their economy was not based on agricultural production. At the time, the scribes of the city of Ur noted that the Amorites were beasts of plunder and did not know of the grain.[17] This would tend to support the idea that the Amorite culture was based on a non-agricultural economy.

While the drought had ended in 1900 BC, leadership of the region had passed from the Akkadians to the city of Ur, and thence onto Babylon, the increasingly dominating city of the region and the home of the Amorites.[18] Ibbi-Sin was the last ruler of Ur, who lost control of Mesopotamia simply because of communication breakdowns with other cities he controlled, who saw it as a chance to reassert their independence.[19] So, although the downfall of the civilisation of Ur may not have been directly due to Amorite military pressure, the state was weakened from within. Furthermore, the dependence of Ur on the use of Amorite mercenaries may explain why Amorite dynasties emerged so rapidly from the ashes of Ur.[20] Also, the arrival of the Amorite dynasties completely replaced the preceding civilisation, in both Mesopotamia and in Egypt.[21] The break between the civilisations was so distinct that it was deemed necessary to also make a distinction between two historical periods.[22] In this case, the time before the Amorites is known as the Early Bronze Age,

---

[14] Ibid.

[15] Ibid.

[16] De Lafayette, M. (2014), Thesaurus Lexicon of Similar Words & Synonyms in 21 Dead and Ancient Languages, Volume II 'A' 9Alma-Azu), Times Square Press, New York.

[17] Wilford, J. (1993), Collapse of Earliest Known Empire Is Linked to Long, Harsh Drought, The New York Times.

[18] Ibid.

[19] Whiting, R. M. (1995), Amorite tribes and nations of second millennium western-Asia. In Civilizations of the Ancient near East, Vol. 2 (ed. J. M. Sasson, with J. Baines, G. Beckman and K. Robinson), New York: Charles Scribner's Sons, pp. 1231–42.

[20] Ibid.

[21] Kenyon, K. (1965), Archaeology in the Holy Land, 2nd edition, Ernest Benn.

[22] Prag, K. (1985), Ancient and Modern Pastoral Migration in the Levant, Levant, 17:1, pp. 81–88.

and the time after they came became known as the Intermediate Early Bronze- Middle Bronze Age (EB, MB).[23] The dispersion of Amorite culture from Syria to Mesopotamia may have started has a gradual process in the 3rd millennium BC, culminating in the early 2nd millennium BC.[24] It is because of this that there has been some debate as to whether the Amorite incursion into Mesopotamia was caused by a military incursion or by a process of economic colonisation.[25] The archaeological evidence is more in favour of the latter, with such evidence also supporting the idea that the Mesopotamian kings militarily campaigned against Amorite settlements as far back as 2250 BC.[26] However, this persecution only helped the Amorites forge a distinct national identity.[27] This would have helped the Amorites become a more cohesive fighting force at the time of the fall of the Sumerian dynasty in 2000 BC, helping them to become dominant in Mesopotamia. The Amorites brought with them new burial techniques as well as new types of pottery.[28] Once the Amorites had become established in the southern Levant, they controlled a number of towns, including Ashkelon, Byblos, Ullaza, and Irqatum.[29] Nevertheless, the archaeological evidence is not so robust as to allow for a detailed historical reconstruction of either the names of the rulers of this period or the span of the territories they controlled.[30] However, historical sources do suggest that, by the end of the intermediate Middle Bronze Age, the whole of the Levant was dominated by Amorite kingdoms.[31] At its heart lay Babylon. However, by 1600 BC, the Amorites of

---

[23] Ibid.

[24] Burke, A. (2014), 'Introduction to the Levant During the Middle Bronze Age', IN: The Oxford Handbook of the Archaeology of the Levant C: 8000–332 BCE, Steiner, M., and Killebrew, A. (Eds), Oxford University Press, Oxford.

[25] Ibid.

[26] Ibid.

[27] Ibid.

[28] Prag, K. (1985), Ancient and Modern Pastoral Migration in the Levant, Levant, 17:1, pp. 81–88.

[29] Burke, A. (2014), 'Introduction to the Levant During the Middle Bronze Age', IN: The Oxford Handbook of the Archaeology of the Levant C: 8000–332 BCE, Steiner, M., and Killebrew, A. (Eds), Oxford University Press, Oxford.

[30] Ibid.

[31] Ibid.

Babylon and the middle Euphrates region were coming under increasing pressure from the Hurrians, Kassites, and the Hittites.[32] It would be the latter who would end the glorious reign of Hammurabi (1728 BC–1686 BC) and his dynasty in Babylon.[33] Of all the kings of the Amorite dynasties of Babylon, Hammurabi was probably the best known historically. The Babylonian Empire vastly expanded once he ascended to the throne. In the first few decades of his rule, Hammurabi focused on expanding public works, such as strengthening the defensive walls of Babylon and expanding the temple network within the city.[34] He also freed Babylon from Elamite dominance, and conquered the cities of Isin, Larsa, Eshnunna, Kish, Lagash, Nippur, Ur, Uruk, Umma, Adab, and Endu.[35] Furthermore, Hammurabi expanded fortifications and temple networks in other parts of the Babylonian Empire, improved irrigation by organising the digging of a canal between Kish and the Persian Gulf; and laid down the Code of Hammurabi.[36] It was also after the reign of Hammurabi that the development of algebra began in around 1650 BC to 1700 BC, which encompasses the period of the reign of Hammurabi, his son, and grandson.[37] However, this development was not unique to Babylonia because such work was also being carried out in Egypt at the same time,[38] but it is uncertain as to whether there was a diffusion of knowledge between Babylonia and Egypt.[39] However, knowledge of algebra did spread from Babylonia to Greece between the fifth and the third centuries BC, Arabia and India in 700 AD, and Europe

[32] Whiting, R. M. (1995), Amorite tribes and nations of second millennium western-Asia. In Civilizations of the Ancient Near East, Vol. 2 (ed. J. M. Sasson, with J. Baines, G. Beckman and K. Robinson), New York: Charles Scribner's Sons, pp. 1231–42.

[33] Ibid.

[34] Sharma, S., and Pahuja, D. (2017), Five Great Civilisations of Ancient World, Educreation Publishing, New Delhi.

[35] Ibid.

[36] Ibid.

[37] Stallings, L. (2000), A Brief History of Algebraic Notation, School of Science and Mathematics, Vol. 100, No. 5, pp. 230–235.

[38] Ibid.

[39] Ibid.

in 1100 AD.[40] Nevertheless, it is known from written narratives such as the Ahmes papyri and Hammurabi-period clay tablets that Egyptian algebra was less sophisticated than Babylonian algebra.[41] This may have been because the numerical system in use in Egypt was less well developed than what was in use in Babylonia at the time.[42] The origins of Mathematics in Babylonia lies with the Sumerians and their development of writing in 3000 BC. It is believed that the Sumerian spoken language of the time utilised a sexagesimal counting structure based on number words.[43] Furthermore, there was a connection between Babylonian place notation and the sexagesimal system of counting.[44] The development of mathematical notations and a system of writing must have developed around the same time during the Sumerian period of Babylonia. Moreover, the development of writing also developed in parallel with the emergence of the cities of ancient Sumer.[45] These ancient cities began to agglomerate around centres of worship such as temple complexes, which also acted as storehouses, and as sites of scholarship and learning.[46] The need to store goods for future consumption would have required a way in which to record the quantities of such goods being kept. This may have been the way by which primitive writing first emerged in the world. Temples also owned the largest estates in a society in which most people survived on a subsistence living.[47] However, in

[40] Baumgart, J. K. (1969), The history of algebra: An overview, In Historical topics for the mathematics classroom. 31st National Council of Teachers of Mathematics Yearbook. Washington, DC: NCTM.

[41] Gullberg, J. (1997), Mathematics from the birth of Numbers, W.W. Norton, New York.

[42] Stallings, L. (2000), A Brief History of Algebraic Notation, School of Science and Mathematics, Vol. 100, No. 5, pp. 230–235.

[43] Powell, M. (1976), The Antecedents of Old Babylonian Place Notation and the Early History of Babylonian Mathematics, Historia Mathematica J, 3, pp. 417–439.

[44] Ibid.

[45] Modelski, G. (1999), Ancient World Cities 4000–1000 BC: Centre/hinterland in the World System, Global Society, 13:4, pp. 383–392.

[46] Ibid.

[47] Jursa, M. (2010), Aspects of the Economic History of Babylonia in the First Millennium BC. Economic Geography, Economic Mentalities, Agriculture, the Use of Money and the Problem of Economic Growth, Ugarit Verlag, Münster.

contrast to Egypt—where the temples and, ultimately, the Pharaoh owned all land—in Babylonia, there was a tradition of the private ownership of land as well as ownership of land by the King and by the temples.[48] Temples may also have been the place from which prostitution emerged into a commercial activity, following sacred prostitution in the context of fertility cults and the worship of goddesses associated with fertility.[49] However, by 1200 BC, a new alphabet-based form of writing had emerged and spread from the coast of the Levant.[50] Moreover, while the development of cities in the near east as well as in the Indus Valley occurred under the same conditions, and Sumer ceased to be the centre of the urban system.[51] It is also interesting to note that the ancient civilisations of Mesopotamia, Egypt, and the Indus Valley were all active at parallel times in history, with the latter being larger than the first two.

Hittites and the Kassites: 1595 BC

Samsu-ditana (1625 BC–1595 BC) was the last of the kings of the first dynasty of Babylon.[52] By 1595 BC, the Babylonian Empire had become so militarily weak that the Hittite king, Mursilis I, raided Babylon unopposed and took away the city's deities, Marduk and his consort.[53] However, Mursilis I did not occupy Babylon, and nothing was heard of Samsu-ditana and the royal family after the raid.[54] But the raid and its

[48] Carmona, S., and Ezzamel, M. (2007), Accounting and accountability in ancient civilisations: Mesopotamia and ancient Egypt, Accounting, Auditing & Accountability Journal, Vol. 20, Issue 2.

[49] Bullough, V and Bullough, B. (1978), The History of Prostitution: An Illustrated Social History, Crown Publishers, New York.

[50] Modelski, G., and Thompson, W. (1999), Evolutionary Pulsations in the World Economy, Paper presented to the International Studies Association (ISA) Annual Convention, San Diego, April 1996, The Evolutionary Pulse of the World System: Hinterland Incursions and Migrations 4000 BC to 1500 AD, in P.N. Kardulias (ed.), World Systems Theory in Practice: Leadership, Production, and Exchange, Rowman and Littlefield, Boulder.

[51] Modelski, G. (1999), Ancient World Cities 4000–1000 BC: Centre/hinterland in the World System, Global Society, 13:4, pp. 383–392.

[52] Saggs, H. (2000), Peoples of the Past: Babylonians, University of California Press, Berkeley.

[53] Ibid.

[54] Ibid.

aftermath may have proved to be an opportune moment for the nearest kingdom, the Sealands to the south of Babylon, to seize the city.[55] This view is evidence by the existence of a boundary stone in the region which names Gulkishar, a king of the Sealands dynasty, as a king of Babylon.[56] Nevertheless, the rule of the Sealands dynasty over Babylon must have been a brief one because Babylon was taken over and ruled by the Kassites, another peoples of the Zagros, from 1595 BC to 1157 BC.[57] The Kassite conquest of the Sealands dynasty had probably been completed by the latter part of the sixteenth century.[58] The defeat of the Sealands allowed the Kassites to create a national identity over southern Mesopotamia.[59] However, while the Kassites instituted a new form of political organisation over southern Mesopotamia, the old city state model being abandoned in favour of a national one, they abandoned hope of reintegrating northern Mesopotamia under their hegemony, settling its rule by the Assyrians through a treaty.[60] The formerly autonomous city states became provincial capitals, with Babylonia itself divided into 20 provinces.[61] Each province was governed by a Governor, appointed by the royal court in Babylon, who would be responsible for a hierarchy of functionaries with distinctly new titles from the previous ancient ones.[62] However, provinces at the frontier, further away from Babylon, were governed along tribal, family lines.[63] This would have allowed for greater control, compared to the case where a royal appointed governor would have less regional influence as well as less local knowledge.[64] The temples in Babylonia gained more influence as administrative centres, with complex

[55] Ibid.

[56] Ibid.

[57] Ibid.

[58] Stiebing, W. (2009), Ancient Near Eastern History and Culture, 2nd Edition, Routledge, London.

[59] Ibid.

[60] Ibid.

[61] Liverani, M. (2011), The Ancient Near East: History, Society and Economy, Routledge, New York.

[62] Ibid.

[63] Ibid.

[64] Ibid.

but logical accounting systems in place.[65] This institutional support allowed for a stable and diversified Kassite economy.[66] In this economy, the Kassites also put in place a local taxation system, province by province, upon which was the basis of the revenue streams for the king and the state.[67] The tax was imposed on crop production and on animal husbandry, and was collected from urban and from rural populations.[68] Land grants and tax exemptions would be given by the king to those in his favour, promoting more agricultural production, and generating more revenues for the king.[69] There would also have been an increase in the use of labour for agricultural production and for animal husbandry, consequently increasing employment levels in the rural and urban sectors.[70] The proximity of the cities of Mesopotamia to the Euphrates and Tigris rivers would have ensured low transport costs, as well as acting as the agent of the economic integration of space.[71] However, the integration of economic space in ancient Mesopotamia did not necessarily also mean that there would be economic homogeneity.[72] In this case, land in different regions was used to grow different crops, and the nature of tenure was also different between regions.[73]

Before, their conquest of Mesopotamia, it is known that the Kassites were active in Babylonia, both militarily, and perhaps even economically, having fought with Samsu-ditana and his son, as well as peacefully migrating to the region and the environs of Babylon itself.[74] It is also possible that a Kassite kingdom had become established, at the time, in the

---

[65] Ibid.

[66] Ibid.

[67] Ibid

[68] Ibid.

[69] Ibid.

[70] Ibid.

[71] Jursa, M. (2010), Aspects of the Economic History of Babylonia in the First Millennium BC. Economic Geography, Economic Mentalities, Agriculture, the Use of Money and the Problem of Economic Growth, Ugarit Verlag, Münster.

[72] Malko, H. (2014), Investigation into the Impacts of Foreign Ruling Elites in Traditional State Societes: The Case of the Kassite State in Babylonia (Iraq), ProQuest, Ann Arbor, Michigan, USA.

[73] Ibid.

[74] Saggs, H. (2000), Peoples of the Past: Babylonians, University of California Press, Berkeley.

middle Euphrates region under Hittite protection.[75] The name of the king of the Khana kingdom, in the middle Euphrates, was Kassite.[76] This would explain why it would have been possible for the Hittite king Mursilis I to pass through and to pass out of the region to raid Babylon, unhindered.[77] It was at the time the Kassites were dominant in Babylon that another ethnic group, the Hurrians, introduced the horse into Mesopotamia.[78] It would also seem that the Hurrians were the first peoples to use the two-wheel, horse-drawn chariot for military purposes.[79] However, it is equally possible that the Kassites from Central Asia were the first to introduce the horse-drawn chariot into Mesopotamia.[80] It is possible that Kassites may have been serving in the Babylonian military under the previous regime, prior to their takeover.[81] This may have given the Kassites an advantage in the takeover of Babylonia when the previous regime collapsed. The technological innovation was that the wheels of the chariots used by the Kassites/Hurrians were spoked, rather than consisting merely of pieces of heavy wood bolted together.[82] The spoked design of the wheels allowed the chariots to be driven faster and to be more manoeuvrable, giving the Kassites an advantage in battle. The horse could have been domesticated in the Ukraine region in the fourth millennium BC, arriving in Anatolia and Iran by the third millennium BC; and in Susa and northern Mesopotamia by the first half of the second millennium BC.[83] The Hurrians acted as a conduit of knowledge from one region to another, not just knowledge of the horse, but also cultural

---

75 Malko, H. (2014), Investigation into the Impacts of Foreign Ruling Elites in Traditional State Societes: The Case of the Kassite State in Babylonia (Iraq), ProQuest, Ann Arbor, Michigan, USA.

76 Ibid.

77 Ibid.

78 Ibid.

79 Ibid.

80 McNeese, T. (2002), Ancient Egypt and other Early Civilisations: The Ancient World Set II, Milliken Publishing Company, a Lorenz Company, Dayton, Ohio, US.

81 Malko, H. (2014), Investigation into the Impacts of Foreign Ruling Elites in Traditional State Societes: The Case of the Kassite State in Babylonia (Iraq), ProQuest, Ann Arbor, Michigan, USA.

82 McNeese, T. (2002), Ancient Egypt and other Early Civilisations: The Ancient World Set II, Milliken Publishing Company, a Lorenz Company, Dayton, Ohio, US.

83 Saggs, H. (2000), Peoples of the Past: Babylonians, University of California Press, Berkeley.

knowledge. In this case, the Hurrians were the conduit of Mesopotamian literature and culture to the archaic Greeks through the Hittites.[84] The Kassites adopted Babylonian Akkadian for speech as well as for writing.[85] However, the Kassites not only adopted the local language, but they also adopted the local customs and religion.[86] The literature written by the Kassite scribes of the time emphasised self-reflection, as well as a hope for stability and peace.[87] It was during the Kassite period of Babylonia that the classic work of literature, the 'Epic of Gilgamesh', was written and the 'Creation Epic' took its classical form.[88] The former emphasises the journey of the human soul through hardship and toil, emerging as the butterfly of wisdom and inner peace.[89] In addition to contributing towards innovation in military technology, and literature, the Kassites also contributed in other ways to Mesopotamian civilisation. For example, in architecture, a new type of adornment was affixed to mud brick buildings.[90] These adornments were represented as figures in moulded bricks.[91] The radial vault technique was also used in the construction of store rooms at the Kassite Palace of Aqar Ouf, although the vaults were lower in pitch than usual.[92] In the case of the measurement of time, a new year would start from the coronation of a new king.[93] New techniques were developed to produce glass, as result glass utensils were in widespread use at the time.[94] Changes in land ownership were marked physically using boundary stones, known as Kudurru stones.[95] The use of the latter

[84] Ibid.

[85] Stiebing, W. (2009), Ancient Near Eastern History and Culture, 2nd Edition, Routledge, London.

[86] Ibid.

[87] Ibid.

[88] Ibid.

[89] Ibid.

[90] Ibid.

[91] Crawford, H. (2015), Ur: The City of the Moon God, Bloomsbury Academic, London.

[92] Baqir, T. (1945), Iraq Government excavations at Aqar Quf, Iraq Supplement. London.

[93] Stiebing, W. (2009), Ancient Near Eastern History and Culture, 2nd Edition, Routledge, London.

[94] Arnold, B. (2004), Who Were the Babylonians?, Society of Biblical Literature, Atlanta.

[95] Ibid.

signified the emergence of a new form of land ownership in Babylonia, the granting of land ownership by the King to members of the administrative and religious elites.[96] This could have been either as a reward or to buy loyalty. Either way, the deed would have acted as an incentive to excel. A notion of the communal ownership of land also came back into vogue under Kassite rule, with sales of land to 'outsiders' having to be approved by the local tribe.[97] Moreover, the family of the seller of the land also had a right of veto.[98] The sale of land to aliens was therefore restricted, allowing land to be held amongst kinfolk.

The Kassite period of rule of Babylonia provided political and economic stability.[99] This in turn provided an environment which was conducive to innovation and to cultural development. Hypothetically, the stability ensured that some of the state's resources could be diverted towards facilitating education and training. People would also have time to think about other things, rather than to think about ways in which economic and political stability could be maintained on a day to day basis. However, the arrival of the 'Sea People', circa 1200 BC signalled the transition from the Bronze Age to the Iron Age, and the end of many of the preceding civilisations such as the Hittites and the Mycenaeans.[100] The 'Sea People' bought with them new techniques of iron production as well as hundreds of years of political and institutional instability.[101] This period of ancient history has many similarities to the political and economic chaos which followed the collapse of the western Roman Empire many centuries later.[102]

---

[96] Liverani, M. (2011), The Ancient Near East: History, Society and Economy, Routledge, New York.

[97] Ellickson, R., and Thorland, C. (1995), Ancient Land Law: Mesopotamia, Egypt, Israel, Chicago-Kent Law Review, Vol. 71, No. 321.

[98] Ibid.

[99] Liverani, M. (2011), The Ancient Near East: History, Society and Economy, Routledge, New York.

[100] Arnold, B. (2004), Who Were the Babylonians?, Society of Biblical Literature, Atlanta.

[101] Ibid.

[102] Ibid.

The Assyrians: 911 BC

The history of the Assyrians, who were prominent in the period 911 BC to 680 BC, can be divided into three phases, the Old, the Middle, and the Late.[103] In the Old period, between the nineteenth and the eighteenth centuries, Assyrian merchants were trading with urban centres on the Anatolian plateau.[104] In fourteenth century BC, the Assyrians enjoyed greater prosperity and power when they were able to remove the Mitannian hold over them.[105] Power in the Middle Assyrian period extended as far as the Euphrates, although volatile, lasting until the death of King Tiglath-pileser I in 1076 BC.[106] After this time, the Assyrians remained in a state of weakness until the ascendancy of King Adad-Nirari II in 911 BC.[107] The reign of this king begins in the Late Assyrian period, in which the state began to regain central control of the original Assyrian heartlands, and also to conquer new territories.[108] The Late Assyrian period lasted until the destruction of Nineveh, the Assyrian capital, in 611 BC.[109] Following Adad-Nirari II, there was a succession of notable Assyrian kings, which included Assurnasirpal II (883 BC to 859 BC), whose reign would mark the beginning of the Assyrian empire.[110] Under this king, there was also an expansion of the Assyrian kingdom to as far as the shores of the Mediterranean, with other weaker contemporary kingdoms forced to pay tribute to the Assyrian King. The revenues flowing from the tributes were used to finance the building of a new capital of Nimrud, with the old capital, Assur, becoming a religious centre.[111] The city was surrounded by a wall that had been built using 70 million

[103] Scarre, C., and Fagan, B. (2016), Ancient Civilisations, 4th edition, Routledge, New York.
[104] Ibid.
[105] Ibid.
[106] Ibid.
[107] Scarre, C., and Fagan, B. (2016), Ancient Civilisations, 4th edition, Routledge, New York.
[108] Ibid.
[109] Ibid.
[110] Ibid.
[111] Ibid.

sundried bricks.[112] A site of prominence within the city was the royal palace, which incorporated a new form of decoration for its walls, in the form of friezes.[113] These friezes were painted in deep colours, as were the ceilings.[114] Shalmaneser III (858 BC to 824 BC) continued the expansion of the Assyrian empire by holding the lands of Syria and the Mediterranean coastal areas.[115] Nevertheless, while the kingdoms to the west of the Euphrates had remained independent from Assyrian control, during the declining years of the rule of Shalmaneser III, they sought to reassert their full independence.[116] However, the tables were turned during the reign of Tiglath-Pileser III (744 BC to 727 BC), when the western kingdoms were conquered and integrated into the Assyrian Empire.[117] The progressive expansion of the Assyrian Empire continued under Sargon II (721 BC to 705 BC) and Sennacherib (704 BC to 680 BC).[118] The latter moved the capital of the Assyrian Empire from Nimrud to Nineveh.[119] And it was Sargon II who razed the city of Babylon to the ground, ending its centuries of dominance over northern Mesopotamia and the region. Following Sennacherib, the subsequent kings of the Assyrian Empire, Esarhaddon (680 BC to 669 BC) and Assurbanipal (669 BC to 627 BC), conquered Egypt and the Elamites in 671 BC and 647 BC respectively.[120] However, Egypt would fall out of the grasp of the Assyrians, and Elam was conquered at a high cost.[121] During Assurbanipal's reign, palace libraries maintained clay tablets which depicted religious, mathematical, scientific, and religious works of scholarship.[122] The King's agents were sent near and far to find ancient and contemporary clay tablets for the

112 Ibid.
113 Ibid.
114 Ibid.
115 Ibid.
116 Ibid.
117 Ibid.
118 Ibid.
119 Scarre, C., and Fagan, B. (2016), Ancient Civilisations, 4th edition, Routledge, New York.
120 Ibid.
121 Ibid.
122 Ibid.

palace library at Nineveh.[123] Around 20,000 clay tablets survived the fires which ravaged the palaces of Nineveh in 612 BC, and these tablets are now housed at the British Museum in London.[124] The libraries of Nineveh would tend to substantiate the fact that the Assyrian empire, at its height, was not just a military machine, but also a centre of learning and scholarship. Specialists and scholars from newly conquered territories were transported to the heart of the Assyrian empire so that it should benefit from a diffusion of knowledge facilitating further innovation.[125] The Late Assyrian empire also saw the growing importance of trade within it, in which Karum's—or trading posts—were established.[126] Furthermore, to ensure that trade could be effectively carried out, the Aramaic language was adopted throughout the empire.[127]And conquered states were tightly integrated with other parts of the empire, with Assyrian advisors being sent to guide local rulers.[128] As a result of trade, the Assyrian king accumulated capital in the form of gold and silver, which resulted from the payment of taxes by traders to the state.[129] Trade was also facilitated by usury as a primitive form of banking, sited in temples such as the Ishtar of Arbela.[130] In this case, merchants and traders would have been able to borrow 'money' or 'capital' at a specified rate of interest to pay for the resources that would be needed to set up a trading mission with other parts of the empire, and with other civilisations.[131] In this case, as the Assyrian royalty and elite became wealthier and wealthier, the greater the quantity of luxury goods and goods of different types, which they would have needed from other civilisations in the Mediterranean, Africa, the

[123] Ibid.

[124] Radner, K. (2015), Ancient Assyria: A Very Short Introduction, Oxford University Press, Oxford.

[125] Ibid.

[126] Hirmis, A. (2018), The Economics of Iraq: Ancient Past to Distant Future, Grosvenor House Publishing, Guildford, Surrey, UK.

[127] Ibid.

[128] Ibid.

[129] Ibid.

[130] Ibid.

[131] Hirmis, A. (2018), The Economics of Iraq: Ancient Past to Distant Future, Grosvenor House Publishing, Guildford, Surrey, UK.

levant, Arabia, and ancient India.[132] In order to achieve trade to this extent, the Assyrians would have had access to the trading networks of the Phoenicians.[133] Thus, trade in ancient Mesopotamia was more significant under the Assyrian Empire than it had been in Babylonia. This would suggest that it was Assyria rather than Babylonia that served as the conduit for the meeting of different cultural, political, religious, and scientific traditions of the time.[134]

The rise of the Assyrian Empire was centred at a time when the Iron Age began in 1000 BC, facilitating a shift from bronze to iron swords.[135] This shift in technology allowed for greater decisiveness in battle, in which the thrust of the sword would deal a mortal blow to the enemy without the need for too much reliance on body contact.[136] The Assyrians also introduced the use of archers on horseback, giving greater range from which to strike the enemy.[137] This would allow for fewer Assyrian casualties, with large numbers of enemy soldiers having already been struck down before encountering the Assyrian infantry. Reed boats were also used to carry archers into battle across rivers and seas.[138] However, the first fleet of Assyrian ships, with two decks of oarsman, was built in 694 BC and used in the military campaign against the kingdom of Elam.[139] In order to waterproof the ships, pitch or asphalt gathered from surface petroleum deposits[140] would have been used as a sealant. Organisational and procedural innovations were also instituted into the military by the Assyrians. This included not only regular maintenance of weapons, but also training programmes for the soldiers to learn different

---

[132] Ibid.

[133] Ibid.

[134] Wiseman, D. (1965), Assyria and Babylonia, Cambridge University Press, Cambridge.

[135] Dunnigan, J. (2000), Wargames Handbook: How to Play and Design Commercial and Professional Wargames, 3rd edition, Writers Club Press, New York.

[136] Ibid.

[137] Ibid.

[138] Gabriel, R. (2002), The Great Armies of Antiquity, Praeger Publishers, Westport, CT, USA.

[139] Radner, K. (2015), Ancient Assyria: A Very Short Introduction, Oxford University Press, Oxford.

[140] Ray, G. (1979), Energy Economics – A Random Walk in History, Energy Economics, 3, pp. 139–143.

techniques, and perhaps even to specialise in one aspect of warfare.[141] Soldiers were also divided into smaller tactical groups.[142] This allowed for greater flexibility in deployment and control. Assyrian soldiers were also equipped with better clothing and footwear. In the case of the latter, the Assyrian innovation was the equipping of soldiers with knee length leather boots, with thick soles and shin pads of thin iron plates sewn into the boots.[143] This kind of footwear allowed soldiers to fight on different types of terrain throughout the year.[144] Battles were the predominant means of engagement for the Assyrian military, while sieges were to be avoided because of the high cost and the lack of effective weaponry. Catapults that could be used to destroy the walls of a city were not innovated and used until the fourth century BC.[145]

Nabopolassar: 612 BC

The kings of Assyria who followed Ashurbanipal Ashur-etil-ilani, Sin-shum-lishir and Sin-shar-ishkun were less militarily and politically skilled and able than the kings they succeeded. This may partly explain why the Assyrian Empire went into decline, unable to survive growing internal dissension. The last king of Assyria was Sin-shar-ishkun, who reigned until 612 BC, when Nineveh was destroyed by a military coalition of the Babylonians, Medes, Scythians, and the Chaldean's.[146] Nabopolassar (625 BC to 605 BC) was the Chaldean king of Babylonia who usurped Assyria as the dominant power in the region by leading the forces which destroyed Nineveh in 612 BC. 'Chaldean' is a term used by the Greeks to

[141] Dunnigan, J. (2000), Wargames Handbook: How to Play and Design Commercial and Professional Wargames, 3rd edition, Writers Club Press, New York.

[142] Gabriel, R. (2002), The Great Armies of Antiquity, Praeger Publishers, Westport, CT, USA.

[143] Ibid.

[144] Ibid.

[145] Radner, K. (2015), Ancient Assyria: A Very Short Introduction, Oxford University Press, Oxford.

[146] Finegan, J. (1969), Light from the Ancient Past: The Archaeological Background of the Hebrew-Christian Religion, Vol. 1, 2nd edition, Princeton University Press, Princeton.

describe the tribes of Kaldu, who had been in occupation of the extreme southern part of Babylonia since the latter half of the ninth century BC.[147] The Chaldeans, like the original inhabitants of Babylonia, worshipped Marduk, the deity of the city of Babylon.[148] This would facilitate the binding of the Chaldeans with the inhabitants of Babylonia, making them essentially one.

There is some dispute as to how Nabopolassar became King of Babylonia. Some say that he simply seized power following the death of Babylonia's Assyrian king Ashurbanipal, who had succeeded his brother Shamash-shum-ukin, as King of Babylonia.[149] On the other hand, others say that Nabopolassar was an Assyrian envoy sent to southern Babylonia who, seeing the chaos, seized the moment to take the throne and become King of Babylon himself.[150] In this case, it would be fair to say that the final years of the Assyrian Empire are shrouded in mystery, with light being shed on this period by the works of Roman and Greek writers 200 years later, as well as by the works of the old Testament prophets.[151] Nevertheless, however it happened, Nabopolassar did become King of Babylonia and, in doing so, he reconstituted a new Babylonian Empire.[152] However, while the capital of the Assyrian Empire, Nineveh, had fallen to Nabopolassar in 612 BC, what was left of the Assyrians all moved to the western city of Haran where an Assyrian King, Ashur-uballit, continued to rule for a few years until 606 BC, when Haran too fell to the Scythians.[153] However, once Haran had fallen, the Assyrians moved their capital to Carchemish, which was on the banks of the Euphrates River. It was at Carchemish that an alliance of Assyrians and Egyptians were

---

[147] Sack, R. (2004), Images of Nebuchadnezzar: The Emergence of a legend, Selinsgrove, Susquehanna University Press, London.

[148] Ibid.

[149] Finegan, J. (1969), Light from the Ancient Past: The Archaeological Background of the Hebrew-Christian Religion, Vol. 1, 2nd edition, Princeton University Press, Princeton.

[150] Gill, A. (2010), Gateway to the Gods: The Rise and Fall of Babylon, Quercus.

[151] Price, I. (1924), The Nabopolassar Chronicle, Journal of the American Oriental Society, Vol. 44, pp. 122–129.

[152] Finegan, J. (1969), Light from the Ancient Past: The Archaeological Background of the Hebrew-Christian Religion, Vol. 1, 2nd edition, Princeton University Press, Princeton.

[153] Ibid.

defeated by a Babylonian army under the command of Nebuchadnezzar, the King of Babylon and the son of Nabopolassar, in 605 BC. There is no evidence to support the idea that there was more than one battle at Carchemish.[154] This defeat removed the Assyrians as a political entity in Mesopotamia, in which Babylonia under the Chaldean's was now making a resurgence, both politically and economically. While the Assyrians had been completely crushed, remnants of the Egyptian army escaped back to Egypt. However, Egypt never regained the power it once had and the Pharaoh Neco (609 BC to 595 BC) had to cede Egyptian territories to Babylon. The combined armies of Egypt and Assyria had been outmatched at the Battle of Carchemish by the armies of the Babylonians, who had formed military alliances, amongst others, with the Medes allowing for greater action in the Tigris and Euphrates regions, to the surprise of the Assyrians.[155] One of the reasons put forward for the demise of the Assyrian empire was that the Assyrians did not respect either Babylonian culture or Babylonian religion, and they were seen by the Babylonians to be morally corrupt.[156] The feeling of the desecration of their religion and their civil society, which the Babylonians must have long held, was underlined when Babylon itself was sacked by the Assyrian King Sennacherib in the seventh century BC.[157] The behaviour of the Assyrians was in contrast to the morally upright and humble faith of Nabopolassar in the Babylonian god Marduk, the deity of the city of Babylon itself. The strength of Nabopolassar's faith may have encouraged the people of Babylon to rally to him, whilst the behaviour of the Assyrians repelled the people of Babylonia. Moreover, the sudden and quick collapse of the Assyrian Empire facilitated the rise of a political, cultural, and economic vacuum which nearby states sought to fill.[158]

[154] Thiele, E. (1956), New Evidence on the Chronology of the Last Kings of Judah, Bulletin of the American Schools of Oriental Research, No, 143, pp. 22–27.

[155] Price, I. (1924), The Nabopolassar Chronicle, Journal of the American Oriental Society, Vol. 44, pp. 122–129.

[156] Beaulieu, P. (2003), Nabopolassar and the Antiquity of Babylon, Eretz-Israel: Archaeological, Historical and Geographical Studies, Vol. Hayim and Miriam Tadmar Volume, pp. 1–9.

[157] Ibid.

[158] Beaulieu, P. (2003), Nabopolassar and the Antiquity of Babylon, Eretz-Israel: Archaeological, Historical and Geographical Studies, Vol. Hayim and Miriam Tadmar Volume, pp. 1–9.

However, this was only accomplished by Babylon itself, under the unifying figure of Nabopolassar, and then his son Nebuchadnezzar. The latter is best known for the conquest of Judea, with the fall of Jerusalem in 586 BC,[159] and the deportation of the Jews, as well as for the building of imposing walls around the city of Babylon.[160] Moreover, Nebuchadnezzar was also the first King of Babylon to simultaneously also rule over Egypt.[161] He is famed for being responsible for the famous 'Hanging Gardens of Babylon'. This was both architecturally and technically an innovation, because the growing of plants upon vaults required a unique irrigation system.[162] It was also known that Nebuchadnezzar was responsible for other large building works. For example, the building of two cross-country walls to keep invaders out of Babylonia.[163] Nebuchadnezzar also continued the Assyrian policy of resettling conquered tribes in Babylonia, and this ensured an increase in the population of Babylonia.[164] It is interesting to note that the Babylonians had no concept of 'race', as exists in the modern world, and neither did their language allow for the articulation of racial distinction.[165] This lack of racial distinction, between themselves and others, made it easier for the Babylonians to deport conquered tribes back to Babylonia. As a result, the population of Babylonia was racially diverse in the modern context.[166] Moreover, the

[159] Horn, S. (1968), The Babylonian Chronicle and the Ancient Calendar of the Kingdom of Judah, Andrews University Seminary Studies 6, pp. 29–45.

[160] Sack, R. (2004), Images of Nebuchadnezzar: The Emergence of a legend, Selinsgrove, Susquehanna University Press, London.

[161] Lambert, W. (1965), Nebuchadnezzar King of Justice, Iraq, Vol. 27, Issue 1, pp. 1–11.

[162] Dalley, S. (1994), Nineveh, Babylon and the Hanging Gardens: Cuneiform and Classical Sources Reconciled, Iraq, Vol. 56, pp. 45–58.

[163] Baker, H. D. (2012). The Neo-Babylonian Empire. In Potts, D. (ed.), A Companion to the Archaeology of the Ancient Near East, Wiley-Blackwell, Oxford, pp. 914–930.

[164] Wunsch, C. (2010), Neo-Babylonian Entrepreneurs, IN: The Invention of Enterprise: Entrepreneurship from Ancient Mesopotamia to Modern Times, Landes, D., Mokyr, J., and Baumol, W. (Eds), Princeton University Press, Princeton.

[165] Bahrani, Z. (2006), Race and ethnicity in Mesopotamian Antiquity, World Archaeology, 38:1, pp. 48–59.

[166] Bertin, G. (1889), The Races of the Babylonian Empire, The Journal of the Anthropological Institute of Great Britain and Ireland, Vol. 18, pp. 104–120.

increase in population created more economic needs, fostering more economic activity and, therefore, facilitating increased economic growth. Economic growth was also facilitated by entrepreneurial activity. In contrast to Roman times, in the time of Nebuchadnezzar, entrepreneurship was not seen in a negative way, but was regarded as a socially rewardable activity.[167] However, it was typically the case that entrepreneurs did not arise from wealthy families.[168] Moreover, entrepreneurs were social climbers without a family name, and they did not arise from the urban establishment.[169] The most successful entrepreneurs, established either by working in the administrative structure of the royalty or by having close links with it.[170] Once established, the entrepreneurs would then be able to climb up the social ladder by marrying into established families.[171] This would allow them to increase their network capacity, and thus bring more financial opportunities.[172] Land ownership and property rights were the basis of entrepreneurship in neo-Babylonian times.[173] The land was the basis of self-sufficiency for the family, and any excess agricultural or animal production could be sold, generating profits. These profits were then used to buy more land, which could be rented out to tenant farmers and sharecroppers.[174] However, entrepreneurial opportunities were also available for transportation of agricultural produce from rural farms to urban centres.[175] Throughout the history of Babylonia, property rights, law, and taxation played a role in economic growth and economic activity in the state. It would follow that, in

[167] Ibid.

[168] Ibid.

[169] Wunsch, C. (2010), Neo-Babylonian Entrepreneurs, IN: The Invention of Enterprise: Entrepreneurship from Ancient Mesopotamia to Modern Times, Landes, D., Mokyr, J., and Baumol, W. (Eds), Princeton University Press, Princeton.

[170] Ibid.

[171] Ibid.

[172] Ibid.

[173] Ibid.

[174] Ibid.

[175] Ibid.

ancient times, when states are competing for the dominance of wealth and political power through military means, such a quest would need a vast amount of resources, financial and material, to be successful. Such a situation would only arise when entrepreneurship was vibrant and unconstrained. In the case of Babylonia, the deported Judeans continued their tradition of entrepreneurship in Babylonia itself, as well as when some returned to Judea.[176] Entrepreneurship is also the engine of institutional development.[177] Such a situation would only be possible if the rulers of the successful states—in the long run—would need to grant private property rights, ensure the rule of law, as well as maintain a fair and effective tax system.[178] That ancient Babylonia was a market economy has been established through the analysis of data relating to how much one Shekel of silver would buy in the context of six goods, barley, dates, mustard, cardamom, sesame, and wool for the period 464 BC to 72 BC. This data was found by archaeologists to be recorded in cuneiform on stone clay tablets in the ancient city of Babylon. The data analysis of the prices suggested a random walk, and therefore that the ancient Babylonian economy, at least in the period 464 BC to 72 BC, was a market economy.[179] This would lead to greater incentives to make profits, leading to the greater availability of resources, which would be able to support more scholarship, learning, and innovation. As can be seen in Babylonia, there was a fair amount of military innovation. However, the level of innovation will have a direct correlation with the level of economic development.[180]

[176] Ackroyd, Peter R. 1968. Exile and Restoration: A Study of Hebrew Thought of the Sixth Century B.C. Philadelphia, Westminster Press.

[177] Abrutyn, S. (2015), The institutional evolution of religion: innovation and entrepreneurship in ancient Israel, Religion, 45:4, 505–531.

[178] Bernholz P. (1998), International Competition Among States: Institutions, Market Regime and Innovations in Antiquity. In: Bernholz P., Streit M.E., Vaubel R. (eds) Political Competition, Innovation and Growth. Springer, Berlin, Heidelberg.

[179] Temin, P. (2002), Price Behaviour in Ancient Babylon, Explorations in Economic History, 39, pp. 46–60.

[180] Bernholz P. (1998), International Competition Among States: Institutions, Market Regime and Innovations in Antiquity. In: Bernholz P., Streit M.E., Vaubel R. (eds) Political Competition, Innovation and Growth. Springer, Berlin, Heidelberg.

Cyrus the Great: 539 BC

The first clash of the Babylonian's with the Persians probably occurred in southern Babylonia in 540 BC.[181] This prompted Nabonidus to request the sending of all of the statues of the gods from temples in the provincial cities of Babylon.[182] However, Cyrus II of Persia conquered Babylon in 539 BC, and this ended the 11th Babylonian dynasty.[183] Allegedly, the Persians and their allies, the Medes, had diverted the flow of the Euphrates River so that their soldiers could march across the river bed and underneath the city walls, at the end of the reign of Nebuchadnezzar's grandson.[184] The Persians left Babylonian institutional and administrative structures in place, so that the economic growth of Babylonia could continue, which would only benefit the Persian state.[185] Furthermore, Cyrus incorporated Babylonia, the lands of the Syrian coast, Palestine, Syria, and regions of Mesopotamia, under the province of Babylon in the Persian Empire, ruled by Persian governors.[186] Babylon would remain under the domination of the Persians for the next 200 years before falling under the yoke of Macedonia and Alexander the Great in 330 BC.[187] During their rule, the Persians, notably under the rule of Darius I (521 BC to 486 BC), instituted some administrative changes.[188] This was particularly in the context of the contribution of military assets by tenured farmers, with some families having to contribute archers, some

[181] Zawadzki, S. (2012), The End of the Neo-Babylonian Empire: New Data Concerning Nabonidus's Order to send the Statutes of Gods to Babylon, Journal of Near Eastern Studies, Vol. 71, No. 1.

[182] Ibid.

[183] Lambert, W. (1965), Nebuchadnezzar King of Justice, Iraq, Vol. 27, Issue 1, pp. 1–11.

[184] Rogers, D. (2008), The Weeping Merchants: The Earth Weeps and Heaven Rejoices, Lulu Enterprises, Morrisville, North Carolina.

[185] Wunsch, C. (2010), Neo-Babylonian Entrepreneurs, IN: The Invention of Enterprise: Entrepreneurship from Ancient Mesopotamia to Modern Times, Landes, D., Mokyr, J., and Baumol, W. (Eds), Princeton University Press, Princeton.

[186] Charpin, D. (1995), The History of Ancient Mesopotamia: An Overview, IN Sasson, J. (Ed), Civilizations of the Ancient Near East, Vol. 2, Simon and Schuster, New York.

[187] Ibid.

[188] Ibid.

families contributing chariots, and others still contributing horses.[189] Tenured land was administered through institutions called 'hatrus', in domains which allowed for them to be separated for fiscal purposes as well as with regards to military and agricultural exploitation.[190] The Persian rule of Babylonia, with a number of revolts by the local population, especially between the death of Cambyses and the start of the reign of Xerves.[191] This King has been seen as the one who laid the roots for the demise of the Persian rule of Babylonia.[192] However, the Persians in Babylonia also used it as a base to lay claim to the Persian Empire, notably Darius II in 423 BC and Cyrus the Younger in 401 BC.[193] Persian rule also saw a decline in the use of cuneiform writing for maintaining records in Babylonia, because it was a cumbersome system to use.[194] This may explain why there were more state records on clay tablets during the eight-year rule of Cambyses, compared to the thirty-two-year rule of Artaxerxes II (404 BC to 359 BC).[195] However, traditional Babylonian families still wrote in cuneiform, as these families sought to maintain daily use of the traditions of Babylon and Uruk, during the Seleucid and Arsacid dynasties.[196] In this way, cuneiform was still in use in Babylonia upon the arrival of Alexander the Great in 330 BC, and it continued to be used until 70 BC.[197]

[189] Charpin, D. (1995), The History of Ancient Mesopotamia: An Overview, IN Sasson, J. (Ed), Civilizations of the Ancient Near East, Vol. 2, Simon and Schuster, New York.

[190] Ibid.

[191] Charpin, D. (1995), The History of Ancient Mesopotamia: An Overview, IN Sasson, J. (Ed), Civilizations of the Ancient Near East, Vol. 2, Simon and Schuster, New York.

[192] Kuhrt, A. (1988), The Achaemenid Empire: A Babylonian Perspective, Proceedings of the Cambridge Philological Society (New Series), Vol. 34.

[193] Charpin, D. (1995), The History of Ancient Mesopotamia: An Overview, IN Sasson, J. (Ed), Civilizations of the Ancient Near East, Vol. 2, Simon and Schuster, New York.

[194] Ibid.

[195] Ibid.

[196] Ibid.

[197] Ibid.

References

Abrutyn, S. (2015), The institutional evolution of religion: innovation and entrepreneurship in ancient Israel, Religion, 45:4, 505–531.

Ackroyd, Peter R. 1968. Exile and Restoration: A Study of Hebrew Thought of the Sixth Century B.C. Philadelphia, Westminster Press.

Arnold, B. (2004), Who Were the Babylonians? Society of Biblical Literature, Atlanta.

Bahrani, Z. (2006), Race and ethnicity in Mesopotamian Antiquity, World Archaeology, 38:1, pp. 48–59.

Baker, H. D. (2012). The Neo-Babylonian Empire. In Potts, D. (ed.), A Companion to the Archaeology of the Ancient Near East, Wiley-Blackwell, Oxford, pp. 914–930.

Baqir, T. (1945), Iraq Government excavations at Aqar Quf, Iraq Supplement. London.

Baumgart, J. K. (1969), The history of algebra: An overview, In Historical topics for the mathematics classroom. 31st National Council of Teachers of Mathematics Yearbook. Washington, DC: NCTM.

Beaulieu, P. (2003), Nabopolassar and the Antiquity of Babylon, Eretz-Israel: Archaeological, Historical and Geographical Studies, Vol. Hayim and Miriam Tadmar Volume, pp. 1–9.

Bernholz P. (1998) International Competition Among States: Institutions, Market Regime and Innovations in Antiquity. In: Bernholz P., Streit M.E., Vaubel R. (eds) Political Competition, Innovation and Growth. Springer, Berlin, Heidelberg.

Bertin, G. (1889), The Races of the Babylonian Empire, The Journal of the Anthropological Institute of Great Britain and Ireland, Vol. 18, pp. 104–120.

Burke, A. (2014), Introduction to the Levant During the Middle Bronze Age', IN: The Oxford Handbook of the Archaeology of the Levant C: 8000–332BCE, Steiner, M., and Killebrew, A. (Eds), Oxford University Press, Oxford.

Bullough, V and Bullough, B. (1978), The History of Prostitution: An Illustrated Social History, Crown Publishers, New York.

Carmona, S., and Ezzamel, M. (2007), Accounting and accountability in ancient civilisations: Mesopotamia and ancient Egypt, Accounting, Auditing & Accountability Journal, Vol. 20, Issue 2.

Charpin, D. (1995), The History of Ancient Mesopotamia: An Overview, IN Sasson, J. (Ed), Civilizations of the Ancient Near East, Vol. 2, Simon and Schuster, New York.

Crawford, H. (2015), Ur: The City of the Moon God, Bloomsbury Academic, London.

Dalley, S. (1994), Nineveh, Babylon and the Hanging Gardens: Cuneiform and Classical Sources Reconciled, Iraq, Vol. 56, pp. 45–58.

De Lafayette, M. (2014), Thesaurus Lexicon of Similar Words & Synonyms in 21 Dead and Ancient Languages, Volume II 'A' 9Alma-Azu), Times Square Press, New York.

Dunnigan, J. (2000), Wargames Handbook: How to Play and Design Commercial and Professional Wargames, 3rd edition, Writers Club Press, New York.

Ellickson, R., and Thorland, C. (1995), Ancient Land Law: Mesopotamia, Egypt, Israel, Chicago-Kent Law Review, Vol. 71, No. 321.

Finegan, J. (1969), Light from the Ancient Past: The Archaeological Background of the Hebrew-Christian Religion, Vol. 1, 2nd edition, Princeton University Press, Princeton.

Gabriel, R. (2002), The Great Armies of Antiquity, Praeger Publishers, Westport, CT, USA.

Gill, A. (2010), Gateway to the Gods: The Rise and Fall of Babylon, Quercus.

Grayson, A. (1980), Assyria and Babylonia, Orientalia, NOVA SERIES, Vol. 49, No. 2, pp. 140–194.

Gullberg, J. (1997), Mathematics from the birth of Numbers, W.W. Norton, New York.

Hirmis, A. (2018), The Economics of Iraq: Ancient Past to Distant Future, Grosvenor House Publishing, Guildford, Surrey, UK.

Horn, S. (1968), The Babylonian Chronicle and the Ancient Calendar of the Kingdom of Judah, Andrews University Seminary Studies 6, pp. 29–45.

Jursa, M. (2010), Aspects of the Economic History of Babylonia in the First Millennium BC. Economic Geography, Economic Mentalities, Agriculture, the Use of Money and the Problem of Economic Growth, Ugarit Verlag, Münster.

Kenyon, K. (1965), Archaeology in the Holy Land, 2nd edition, Ernest Benn.

Kuhrt, A. (1988), The Achaemenid Empire: A Babylonian Perspective, Proceedings of the Cambridge Philological Society (New Series), Vol. 34.

Lambert, W. (1965), Nebuchadnezzar King of Justice, Iraq, Vol. 27, Issue 1, pp. 1–11.

Liverani, M. (2011), The Ancient Near East: History, Society and Economy, Routledge, New York.

Malko, H. (2014), Investigation into the Impacts of Foreign Ruling Elites in Traditional State Societies: The Case of the Kassite State in Babylonia (Iraq), ProQuest, Ann Arbor, Michigan, USA.

McNeese, T. (2002), Ancient Egypt and other Early Civilisations: The Ancient World Set II, Milliken Publishing Company, a Lorenz Company, Dayton, Ohio, US.

Modelski, G. (1999), Ancient World Cities 4000–1000 BC: Centre/hinterland in the World System, Global Society, 13:4, pp. 383–392.

Modelski, G., and Thompson, W. (1999), Evolutionary Pulsations in the World Economy, Paper presented to the International Studies Association (ISA) Annual Convention, San Diego, April 1996, The Evolutionary Pulse of the World System: Hinterland Incursions and Migrations 4000 BC to 1500 AD, in P.N. Kardulias (ed.), World Systems Theory in Practice: Leadership, Production, and Exchange, Rowman and Littlefield, Boulder.

Powell, M. (1976), The Antecedents of Old Babylonian Place Notation and the Early History of Babylonian Mathematics, Historia Mathematica J, 3, pp. 417–439.

Prag, K. (1985), Ancient and Modern Pastoral Migration in the Levant, Levant, 17:1, pp. 81–88.

Price, I. (1924), The Nabopolassar Chronicle, Journal of the American Oriental Society, Vol. 44, pp. 122–129.

Radner, K. (2015), Ancient Assyria: A Very Short Introduction, Oxford University Press, Oxford.

Ray, G. (1979), Energy Economics – A Random Walk in History, Energy Economics, 3, pp. 139–143.

Rogers, D. (2008), The Weeping Merchants: The Earth Weeps and Heaven Rejoices, Lulu Enterprises, Morrisville, North Carolina.

Sack, R. (2004), Images of Nebuchadnezzar: The Emergence of a legend, Selinsgrove, Susquehanna University Press, London.

Saggs, H. (2000), Peoples of the Past: Babylonians, University of California Press, Berkeley.

Scarre, C., and Fagan, B. (2016), Ancient Civilisations, 4th edition, Routledge, New York.

Sharma, S., and Pahuja, D. (2017), Five Great Civilisations of Ancient World, Educreation Publishing, New Delhi.

Stiebing, W. (2009), Ancient Near Eastern History and Culture, 2nd Edition, Routledge, London.
Stallings, L. (2000), A Brief History of Algebraic Notation, School of Science and Mathematics, Vol. 100, No. 5, pp. 230–235.
Temin, P. (2002), Price Behaviour in Ancient Babylon, Explorations in Economic History, 39, pp. 46–60.
Thiele, E. (1956), New Evidence on the Chronology of the Last Kings of Judah, Bulletin of the American Schools of Oriental Research, No, 143, pp. 22–27.
Weiss, H., Courty, M., Wetterstrom, W., Guichard, F., Senior, L., Meadow, R., and Curnow, A. (1993), The Genesis and Collapse of the Third Millennium North Mesopotamian Civilisation, Science, New Series, Vol. 261, No. 5124, pp. 995–1004.
Whiting, R. M. (1995), Amorite tribes and nations of second millennium western-Asia. In Civilizations of the Ancient Near East, Vol. 2 (ed. J. M. Sasson, with J. Baines, G. Beckman and K. Robinson), New York: Charles Scribner's Sons, pp. 1231–42.
Wilford, J. (1993), Collapse of Earliest Known Empire Is Linked to Long, Harsh Drought, The New York Times.
Wiseman, D. (1965), Assyria and Babylonia, Cambridge University Press, Cambridge.
Wunsch, C. (2010), Neo-Babylonian Entrepreneurs, IN: The Invention of Enterprise: Entrepreneurship from Ancient Mesopotamia to Modern Times, Landes, D., Mokyr, J., and Baumol, W. (Eds), Princeton University Press, Princeton.
Zawadzki, S. (2012), The End of the Neo-Babylonian Empire: New Data Concerning Nabonidus's Order to send the Statutes of Gods to Babylon, Journal of Near Eastern Studies, Vol. 71, No. 1.

5

Ancient Greece: 1100 BC to 30 BC

The Dark Ages of Greece (1100 BC to 750 BC)

The collapse of the Mycenaean civilisation of the Mediterranean coincided with the commencement of the Greek Dark Ages.[1] The period of this age was between the thirteenth century BC and the eleventh century BC, and it coincided with a dramatic drop in Greece's population.[2] This drop may have been caused by a fall in agricultural productivity, due to the effects of climate change, being unable to support a large population base.[3] The specific climate change could have been a prolonged drought, leading the soil to become unable to sustain agricultural crops.[4] However, an alternative argument for the collapse of civilisation

[1] Bryant, J. (1996), Moral Codes and Social Structure in Ancient Greece: A Sociology of Greek Ethics from Homer to the Epicureans and Stoics, State University of New York Press, Albany.

[2] Drake, B. (2012), The Influence of Climatic Change on the Late Bronze Age Collapse and the Greek Dark Ages, Journalof Archaeological Science, 39, pp. 1862–1870.

[3] Ibid.

[4] Carpenter, R. (1966), Discontinuity in Greek Civilization. Cambridge University Press, Cambridge, UK.

S. Ramesh, *The Rise of Empires*, https://doi.org/10.1007/978-3-030-01608-1_5

in the eastern Mediterranean between the thirteenth and the eleventh centuries is that there was a collapse of the economic system.[5]Despite what factors may have caused the Mycenaean civilisation to collapse, the next four centuries were characterised by a decline in long distance trade; and the predominance of rural as opposed to urban settlements.[6] These two characteristics would suggest that, during these four centuries, the political and institutional structure of Greek society has also collapsed. This is because it is centralised political control of society with strong institutional support that is capable of facilitating the rise of urban centres, as well as long distance trade to support the increasing populations of these urban centres. In addition to the four centuries spawning rural and self-sustaining agricultural communities with little if any long distance trade, there was also an increase in illiteracy. The lack of political control, the lack of institutions, and the lack of commerce meant that there was no need to keep records or engage in scholarly ways of life. As a result, writing systems such as Linear B disappeared.[7] Furthermore, during the Mycenaean era, there was contact between the Mycenaeans and the Hittites.[8] In order to facilitate dialogue for trade or peaceful coexistence, the Mycenaeans may have been bilingual, speaking their own language as well as the language of the Hittites, Hatti.[9] In any case, once the Mycenaean civilisation collapsed, the use of Hatti in the Aegean and the eastern Mediterranean disappeared.[10] The level of literacy in the Aegean and in the eastern Mediterranean did not begin to increase again until the start of the eighth century BC, with the introduction of the

[5] Drake, B. (2012), The Influence of Climatic Change on the Late Bronze Age Collapse and the Greek Dark Ages, Journal of Archaeological Science, 39, pp. 1862–1870.

[6] Ibid.

[7] Palaima, T.G. (2010), Linear B. In: Cline, E. (Ed.), The Oxford Handbook of the Bronze Age Aegean. Oxford University Press, Oxford, pp. 356–372.

[8] Castleden, R. (2005), Mycenaeans, Routledge, New York.

[9] Janse, M. (2002), Aspects of Bilingualism in the History of the Greek Language, IN: Bilingualism in Ancient Society: Language Contact and the Written Text, Adams, J., Janse, M., and Swain, S. (Eds), Oxford University Press, Oxford.

[10] Fortson IV, B.W. (2004), Indo-European Language and Culture, an Introduction, Blackwell Publishing, Malden, MA.

Phoenician script by a resurgent Greek civilisation.[11] Associated with the loss of political control, supportive institutions, as well as the rise of literacy, it was at also at the end of the Late Bronze Age that urban centres throughout the eastern Mediterranean region were abandoned.[12] However, this abandonment did not occur at the same time, but over the course of two centuries.[13] However, the de-urbanisation process had been completed by 1050 BC.[14] Non-existent civil governance, non-existent political control, and the rise of rural settlements meant that there were no more grand building projects. The result was that the four centuries encompassing the Greek Dark Ages left no architectural heritage.[15] A lack of adequate defences caused by the decline of the Mycenaen civilisation allowed large migrations of the Phoenicians into the eastern Mediterranean.[16]

Archaic Greece (750 BC to 500 BC)

The designation of the period from 750 BC to 500 BC as archaic in the context of ancient Greece is due principally to the form of art that emerged during that period.[17] However, perhaps more importantly, from an economic perspective, is the emergence of a different form of social and political organisation which had never been seen in antiquity.[18] This may have provided the context for the spectacular achievements in Greece's Classical

[11] Sass, B. (2005), The Alphabet at the Turn of the Millennium: The West Semitic Alphabet CA. 1150–850 BCE, the Antiquity of the Arabian, Greek, and Phrygian Alphabets, Ravgon, Tel Aviv.

[12] Mazar, A. (1990), Archaeology of the Land of the Bible 10,000–586 BCE. Doubleday, New York, NY.

[13] Drake, B. (2012), The Influence of Climatic Change on the Late Bronze Age Collapse and the Greek Dark Ages, Journal of Archaeological Science, 39, pp. 1862–1870.

[14] Ibid.

[15] Desborough, V.R. (1964), The Last Mycenaeans and Their Successors. Oxford University Press, Oxford.

[16] Sandars, N.K. (1987), The Sea Peoples: Warriors of the Ancient Mediterranean. Thames and Hudson, New York.

[17] Martin, T. (2013), Ancient Greece – From Prehistoric to Hellenistic Times, Yale University Press, New Haven.

[18] Ibid.

Age, which followed its Archaic Age.[19] However, while democracy bloomed in the Classical Age of Greece, in its Archaic Age, aristocratic rule can be replaced by a narrower governing institution, the tyrant.[20] While all Polis entered the Archaic Age with power concentrated in the hands of a small number of noble families, the Polis exited the age with governance in the context of a broad range of representation by the people themselves.[21] In the Archaic Age, there were hundreds of Polis in ancient Greece and, in each of these, democracy evolved at a different pace.[22] Each Polis was a community of citizens, into which new citizens were readily accepted.[23] In essence, each Polis represented a society which was of a complex hierarchy in nature, comprised of thousands of individual peasant households who owned the means of production.[24] These households neither paid dues to a central government, nor did they rely on a central government for the necessities of life.[25] The concentration of households in the Polis also gave way to an impersonal exchange of goods, shoes, clothes for agricultural products, which over time gave rise to marketised exchange, especially in Athens.[26] However, over the Archaic Age as a whole, a period characterised by increasing population and economic growth, the governance of the Polis was becoming increasingly autarkic with the rise of the tyrant.[27] The tyrant arose in the Polis, which were centres of commercial activity such as Athens, Corinth and Argos, rather than the Polis which were centres of agriculture such as Thebes and Sparta.[28]

---

[19] Snodgrass, A. (1981), Archaic Greece, The Age of Experiment, University of California Press, Berkeley and Los Angeles.

[20] Fleck, R., and Hanssen, F. (2013), How Tyranny Paved the Way to Democracy: The Democratic Transition in Ancient Greece, The Journal of Law and Economics, Vol. 56, No. 2, pp. 389–416.

[21] Grant, M. (1987), The Rise of the Greeks. New York: Charles Scribner.

[22] Fleck, R., and Hanssen, F. (2013), How Tyranny Paved the Way to Democracy: The Democratic Transition in Ancient Greece, The Journal of Law and Economics, Vol. 56, No. 2, pp. 389–416.

[23] Ehrenberg, V. (1937), When did the polis rise?, JHS 57, 147–59.

[24] Morris, I. (1990), The Early Polis as City and State, IN City and Country in the Ancient World, Rich, J., and Wallace-Hadrill, A. (Eds), Routledge, London.

[25] Ibid.

[26] Ibid.

[27] Andrewes, A. (1956), The Greek Tyrants, Hutchinson's University Library, London.

[28] Fleck, R., and Hanssen, F. (2013), How Tyranny Paved the Way to Democracy: The Democratic Transition in Ancient Greece, The Journal of Law and Economics, Vol. 56, No. 2, pp. 389–416.

Tyrants were also in favour of supply-side economic policies, such as the investment in infrastructure, the building of aqueducts, pump houses, and ports, for example.[29] Furthermore, it was in the Polis, in which the tyrants had been in power during the Archaic Age, that democracy flourished during the subsequent Classical Age.[30] Lastly, the end of tyranny always took two forms: either death or exile.[31] Throughout history, tyrants have been seen from a negative point of view. However, in the context of ancient Greece, tyrants were a break from the past, and their succession to power neither depended on parentage or status of birth.[32] One argument that may support the evolution of democracy from tyranny, relates to the supposition that the pro-economic growth policies of the tyrants allowed the citizenry to become economically and financially self-sufficient, such that they entered the next level of individualism, the quest for political freedom. This was in a society in which a measure of a person's personal quality was the level of their economic success.[33]

The evolution of the political and social organisation which culminated in the Archaic age began in ancient Greece's Dark Age.[34] The concern with the political, social, and economic issues were reflected in the poetry of the time.[35] It can be seen as a structural transformation that laid the economic basis for Greek society in the ensuing centuries.[36] However, the change from the Archaic Age to the Classical Age of Greece was more associated with an intellectual transformation.[37] In the structural transformation, urban centres and adjoining regions with villages were organised

[29] Jeffrey, L. (1976), Archaic Greece: The City-States c. 700–500 B.C, St. Martin's Press, New York.

[30] Fleck, R., and Hanssen, F. (2013), How Tyranny Paved the Way to Democracy: The Democratic Transition in Ancient Greece, The Journal of Law and Economics, Vol. 56, No. 2, pp. 389–416.

[31] Ibid.

[32] Saxonhouse, A. (1988), The Tyranny of Reason in the World of the Polis, American Political Science review, Vol. 82, No. 4.

[33] Ulf, C. (2009), The World of Homer and Hesiod, IN A Companion to Archaic Greece, Raaflaub, K., and Van Wees, H. (Eds), Blackwell Publishing Ltd.

[34] Snodgrass, A. (1981), Archaic Greece, The Age of Experiment, University of California Press, Berkeley and Los Angeles.

[35] Kurke, L. (1992), The Politics of Archaic Greece, Classical Antiquity, Vol. 11, No. 1, pp. 91–120.

[36] Ibid.

[37] Ibid.

into city states, or Polis, whose free men, women, and children were classified as citizens, while each city states population would also have included free non-citizen foreigners and non-free slaves.[38] However, it was only the men who had the right to participate in political activities, even though both free men and women were classed as citizens of the Polis, from which is derived the term 'politics'.[39] One of the most important characteristics of these Greek city states of the Archaic Age was the ability of the poorest in society to not only exist as citizens, but also to have a fair degree of political rights as well.[40] While in contemporary democratic societies this may not seem as a privilege but a right, it is astonishing that it existed in antiquity. For example, besides in ancient Rome, which will be examined in the next chapter, it was not recognised until the American Declaration of Independence was made in the late eighteenth century, that all men were created equal. Thus, it is not clear how the poor free citizens of Archaic Greece could be given a certain level of inalienable rights when amongst them existed slaves. In this case, despite the hardships that the poor free of Archaic Greece faced, as did the slaves, the condition they enjoyed of citizenship and limited political rights gave their existence some meaning in contrast to the existence of the slaves.[41] How this new form of social and political organisation began to emerge in the Dark Ages of ancient Greece has been a subject of debate amongst scholars. It may be the case that it may have existed in the cities of Cyprus and Phoenicia, and knowledge of its existence diffused through to the Greeks through contact with near-eastern traders and/or scholars.[42] However, while it may be easier to accept that technological, religious, and cultural innovations could pass from one society to another in this way, innovations in governance would have required equitable power in society. In other words, everyone must have been equal to some extent anyway without a polarisation of power in society. In this case, as the

[38] Martin, T. (2013), Ancient Greece – From Prehistoric to Hellenistic Times, Yale University Press, New Haven.

[39] Ibid.

[40] Ibid.

[41] Ibid.

[42] Ibid.

extinction of the Mycenaean civilisation in the Mediterranean left a void of power, it may be the case that smaller states arose in the context of equitable power.[43] In other words, everyone's contribution to that society was required for survival, so everyone was treated equally. Furthermore, it is inconceivable that the idea of equality or even the idea of citizenship would have come from the states of the near east, where there was a polarisation of power. Moreover, the reinvention of politics in the Greece of the Dark Ages was leading to concepts which had previously been unknown, such as the equality of treatment under the law and the ability to speak one's mind on political issues.[44] This contrasts with the differentiation between the rich and the poor which had been marked in the era of the near east and Greece during the Mycenae civilisation; and was gaining in strength again during the latter part of the Dark Ages.[45] The only sign of class struggle was evidenced by farmers in Attica openly challenging exploitative landlords in the late seventh and the early sixth century BC.[46] However, despite the rising inequality of income in Greek society and rising equality of treatment under the law, in the case of women, there were added restrictions and regulations in the context of property ownership and sexual behaviour.[47]

At the start of the Archaic Age, the Greeks also started to migrate and establish settlements and trading posts in Ionia, southern France, Sicily, and Spain, where they interacted with the Phoenicians at sites where they too had established trading posts and settlements interacting with the indigenous populations.[48] It was in around 750 BC that the island of Thera was occupied and recolonised by the Spartans specifically because Sparta was becoming overpopulated.[49] However, whereas individuals

[43] Martin, T. (2013), Ancient Greece – From Prehistoric to Hellenistic Times, Yale University Press, New Haven.

[44] Ibid.

[45] Ibid.

[46] Rose, P. (2012), Class in Archaic Greece, Cambridge University Press, Cambridge.

[47] Martin, T. (2013), Ancient Greece – From Prehistoric to Hellenistic Times, Yale University Press, New Haven.

[48] Ibid.

[49] Percy, W. (1996), Pederasty and Pedagogy in Archaic Greece, University of Illinois Press, Chicago.

and merchants may have had ships, it is unlikely that, at a community level, the Polis would have possessed a navy.[50] The political, social, and economic disruptions of the Dark Ages as well as the depopulation of Greece, which had occurred at the time, had led to a large amount of available unoccupied land. However, as the population levels began to rise towards the later Dark Ages and into the Archaic Age—a population explosion in the eighth century BC.[51] In fact, the increase in population between 780 BC and 720 BC may have been by a factor of seven.[52] This led to less and less land being available for occupation, so people had to migrate to make a living either through agriculture or through trade, or through a combination of both.[53] The other factor, which led to increasingly less agricultural food production, was that the sons inherited equally after their father's death.[54] The impact of this would have been that, over generations, the size of the plots of land owned by families would become smaller and smaller with the result that not enough food would have been produced to sustain the families.[55] In this case, landowners would have resorted to borrowing grain, as Greece was a premonetary society at the time, in order to feed themselves and their families using themselves or family members as collateral. Furthermore, the grain shortage was exacerbated by big landowners exporting grain to other cities and regions to obtain goods which were not available in the Polis.[56] In this case, these factors associated with a rising population ensured the occurrence of a severe political, social, and economic crisis.[57]

---

[50] De Souza, P. (1998), Towards thalassocracy? Archaic Greek naval developments, IN Archaic Greece: New Approaches and New Evidence, Fisher, N., and Van Wees, H. (eds), Duckworth with The Classical Press of Wales, London.

[51] Snodgrass, A. (1981), Archaic Greece, The Age of Experiment, University of California Press, Berkeley and Los Angeles.

[52] Ibid.

[53] Martin, T. (2013), Ancient Greece – From Prehistoric to Hellenistic Times, Yale University Press, New Haven.

[54] Parker, V. (2007), Tyrants and Lawgivers, IN The Cambridge Companion to Archaic Greece, Shapiro, H. (Ed), Cambridge University Press, Cambridge.

[55] Ibid.

[56] Parker, V. (2007), Tyrants and Lawgivers, IN The Cambridge Companion to Archaic Greece, Shapiro, H. (Ed), Cambridge University Press, Cambridge.

[57] Ibid.

Due to this, Solon was appointed as the second lawgiver in the Polis of Athens in 594 BC.[58] The role of lawgiver developed from that of arbitrator, a position filled by a wise man who would be just and able to mediate between disputing citizenry.[59] Existing laws could be used by aristocratic judges to cause justice and oppression.[60] At the same time there was a requirement for existing laws, regulations, and customs to be reformed to ensure equality of just treatment under the law. Therefore, in order to prevent injustice, to reform existing laws, and to implement new laws, the arbitrator became the lawgiver.[61] The shift towards written laws in archaic Greece can be seen as a shift towards citizen equality and equal treatment under the law.[62] However, the laws were implemented on an ad hoc basis when existing regulations were not able to resolve developing crises.[63] In this case, the development of written laws in archaic Greece should be seen as helping to develop the Polis, rather than as a means of democracy in itself. Nevertheless, the development of written laws can be seen as an important step towards that end. Moreover, the coherency of culture, customs, religious practices, and laws only came about in the archaic Greece of seventh century BC because that is when the laws were written.[64] These written laws, however, are distinct from rules and customs that were orally transmitted.[65] The difference between the two is that the former represents a permanent institution in the context of the Polis, which could not be forgotten.[66] Solon stipulated

[58] Ibid.

[59] Holkeskamp, K. (1992), Written Law in Archaic Greece, Proceedings of the Cambridge Philological Society, 38, pp. 87–117.

[60] Ibid.

[61] Ibid.

[62] Robinson, E. (1997), The First Democracies: Early Popular Government Outside Athens, Franz Steiner Verlag, Stuttgart.

[63] Ibid.

[64] Blok, J. (2018), Retracing Steps: Finding Ways into Archaic Greek Citizenship, IN Defining Citizenship in Archaic Greece, Duplouy, A., and Brock, R. (Eds), Oxford University Press, Oxford.

[65] Gagarin, M. (2003), Letters of the Law: Written Texts in Archaic Greek Law, IN: Written Texts and the Rise of Literate Culture in Ancient Greece, Yunis, H.(Eds), Cambridge University Press, Cambridge.

[66] Ibid.

policies which the Polis of Athens should follow, rather than just specific laws. For example, he forbade the export from Athens of any agricultural produce except olive oil.[67] Furthermore, Solon also required that fathers should teach their sons a trade, the produce of which could be exchanged for agricultural produce so that they could feed themselves in the event of their families not being able to grow enough food.[68] During his tenure, Solon also contributed to Athens' democratic future, through tyranny, by laying its foundations through institutional innovation in a number of ways.[69] Firstly, Solon introduced a class system based on wealth, which in terms of highest to the lowest, was represented by the Pentakosiomedimni, the Hippeis, the Zeugitae, and the Thetes.[70] This classification would also help to make tax collection more efficient. Secondly, Solon established institutions such as the popular law courts, the assembly of the citizens that discussed and formulated policies, and the Council of the Four Hundred.[71] The latter was to organise and present the business to be discussed and considered by the people's assembly.[72] The Council of the Four Hundred, had its members, selected in groups of one-hundred from the several tribes that made up the population of the Polis of Athens.[73] Its other obligation was to appoint the officers of the state.[74] Solon also stipulated regulations for office, the holding of which depended on wealth.[75]

[67] Parker, V. (2007), Tyrants and Lawgivers, IN The Cambridge Companion to Archaic Greece, Shapiro, H. (Ed), Cambridge University Press, Cambridge.

[68] Ibid.

[69] Salmon, J. (2003), Lopping off the heads? Tyranny, politics and the polls, IN The Development of the Polis in Archaic Greece, Mitchell, L., and Rhodes, P. (Eds), Routledge.

[70] Grote, G. (1856), History of Greece, Murray.

[71] Hammond, B. (1895), Political Institutions and the Greeks, Cambridge University Press, Cambridge.

[72] Sealey, R. (1976), A History of the Greek City States, 700–338 BC, University of California Press, Los Angeles.

[73] Ibid.

[74] Sandys, J. (1912), Aristolelous, Athenaian, Politeia, The Lawbook Exchange Ltd., Union, New Jersey.

[75] Hammond, B. (1895), Political Institutions and the Greeks, Cambridge University Press, Cambridge.

The economy of the Polis cannot be described as a capitalistic one.[76] However, while money can be associated with capitalism, coinage did come to be used in Archaic Greece, either in the sixth century BC[77] or at the start of the eighth century BC.[78] Despite the emergence of coinage in Archaic Greece, it is important not to overstate the contribution of coinage to the economy of the Polis.[79] This is because the coins' values represented very large denominations, and their usage was limited to the environs of the Polis.[80] In this case, the only political interest in economic problems emanated from the citizens of the Polis with their daily needs, in terms of food and services.[81] The citizens were seen primarily as consumers and not as producers, the role of which was taken up by the slaves and by the foreign residents of the Polis.[82] This would suggest that the mode of production in ancient Greece should be looked upon from a political perspective rather than from either a capital or kin perspective[83]; from a political perspective, because production depended on status in the context of either being a citizen or being free. Nevertheless, the rapid increase in population and the establishment of new settlements and trading posts may have lent itself to an ever-increasing need for a more complex form of structural organisation for Greek society.[84] Without an effective co-ordination mechanism, the ever-increasing diffusion of knowledge, and the division of labour would have proved to be difficult to manage.[85] In this light, perhaps the equality of treatment and citizenship mechanism which had been evolving since the Dark Ages was the

---

[76] Morris, I. (1986), Gift and Commodity in Archaic Greece, Man, New Series, Vol. 21, No. 1, pp. 1–17.

[77] Kraay, C. (1976), Archaic and Classical Greek coins. London.

[78] Kagan, D. (1982), The dates of the earliest coins, Am. J. Archaeol. 86, 52–73.

[79] Morris, I. (1986), Gift and Commodity in Archaic Greece, Man, New Series, Vol. 21, No. 1, pp. 1–17.

[80] Cook, R. M. (1958), Speculations on the origin of coinage, Historia 7, 257–62.

[81] Ibid.

[82] Ibid.

[83] Ibid.

[84] Snodgrass, A. (1981), Archaic Greece, The Age of Experiment, University of California Press, Berkeley and Los Angeles.

[85] Ibid.

best solution such that it's use was consolidated during the Archaic Age. Citizenship may have been the mechanism for managing interactions within the Polis, but ultimately, it was these interactions which facilitated economic growth. However, having said this, it is evident that economic growth and living standards were higher and bigger towards the end of the Classical Age and the start of the Hellenistic Age of Greece.[86]

Classical Greece (500 BC to 324 BC)

The Classical Age of Greece encompasses the end of the wars with Persia in 479/8 BC and the death of Alexander the Great in 323 BC.[87] The naval battle at Salamis in 480 BC was a turning point in ancient Greek history and, in hindsight, it not only saved Greece, but also western civilisation, from the Persian conquest.[88] This was also a turning point for Greek society because it was the rowers of Athens who powered the Greek ships to victory over the Persians. They represented the poorest of Athenian society, but their contribution to the Greek victory gave them more confidence and a growing interest in participating in the political affairs of the Polis.[89] The conflict with Persia finally came to an end in 479 BC, when the Greeks defeated the Persian army at the Battle of Plataea.[90] This victory ensured that Athens was now a great power; and its military supremacy necessitated that now there was more to discuss in the assembly.[91] However, it was military action by Sparta in 511 BC which dealt a deadly blow to tyranny in Athens and brought about its demise.[92]

---

[86] Morris, I. (2004), Economic Growth in Ancient Greece, Journal of Institutional and Theoretical Economics, Vol. 160, No. 4, pp. 709–742.

[87] Walter, U. (2010), The Classical Age as a Historical Epoch, IN A Companion to the Classical Greek World, Kinzl, K. (Ed), Blackwell Publishing Ltd., Oxford.

[88] Strauss, B. (2004), The Battle of Salamis: The Naval Encounter That Saved Greece and Western Civilisation, Simon & Schuster Paperbacks, New York.

[89] Walter, U. (2010), The Classical Age as a Historical Epoch, IN A Companion to the Classical Greek World, Kinzl, K. (Ed), Blackwell Publishing Ltd., Oxford.

[90] Shepherd, W. (2012), Plataea 479 BC: The Most Glorious Victory Ever Seen, Osprey Publishing.

[91] Walter, U. (2010), The Classical Age as a Historical Epoch, IN A Companion to the Classical Greek World, Kinzl, K. (Ed), Blackwell Publishing Ltd., Oxford.

[92] Osborne, R. (1996), Greece in the Making, 1200–479 BC, Routledge, New York.

It was after this that the Polis of Athens became an egalitarian democracy in which any citizen could make an equal contribution to the city's political discourse.[93] Nevertheless, the rise of an egalitarian democracy was not a smooth one. It was only possible because of institutional-structural, and physical reforms that had been implemented by Solon and Kleisthenes respectively.[94] Much has been discussed about the contributions of Solon to the emergence of Greek democracy in the previous section, but not much has been said about the contribution of Kleisthenes to this. In this case, it was felt at the time that the one constraint to efficiency in Athenian politics was the centuries-old division of its people into the four tribes. Tribal divisions in the assembly would mean that the time taken to discuss and make decisions would be long, if at all possible. Therefore, in order to allow Solon's structural institutional reforms to work, physical reforms of the old tribal groupings were required to ensure that Athenians acted as one nation.[95] In this case, at the start of the Classical Age, Kleisthenes sub-divided the four original tribal groups into ten 'phyle', without abolishing the tribes.[96] Furthermore, the neighbourhoods of the Polis were organised into smaller hamlets, each with its own local assembly and leader.[97] This spread of power along a more diverse base of the population allowed for each citizen to have a say in the political, military, and economic affairs of the state. Each sub-division of land was known as a deme, and its inception and implementation by Kleisthenes was nothing other than a political revolution.[98] It may have emerged as a result of the political infighting of the three main political groupings in Athens of the time.[99] These political groupings were demarcated by geographical features and included those living in the Attica coastal areas, the plains

---

[93] Walter, U. (2010), The Classical Age as a Historical Epoch, IN A Companion to the Classical Greek World, Kinzl, K. (Ed), Blackwell Publishing Ltd., Oxford.

[94] Ibid.

[95] Andrewes, A. (1977), Kleisthenes' Reform Bill, CQ 27, pp. 246–7.

[96] Ibid.

[97] Ibid.

[98] Osborne, R. (1996), Greece in the Making, 1200–479 BC, Routledge, New York.

[99] Gaeta, L. (2004) Athenian democracy and the political foundation of space, Planning Theory & amp; Practice, 5:4, pp. 471–483.

surrounding the Polis of Athens, and the mountainous regions.[100] The Paralion, the Pedieis, and the Hyperakrioi respectively.[101]

The physical reorganisation of the neighbourhoods of the Polis of Athens into demes had a number of repercussions for Athenian politics. These included the remodelling of the army, and the emergence of a new institution of the five-hundred to which each Phlae contributed fifty citizens.[102] Every deme was also represented in this new council.[103] The reforms instituted by Kleisthenes focused on the spatial re-orientation of power.[104] The mechanism of power that had held sway for centuries, that of the aristocratic families through their wealth and bloodlines, was now irrevocably broken.[105] The mechanism of power had now moved from the aristocratic families of the Polis of Athens to the land and its people.[106] This was the first time, in known and recorded human history that this had happened and, many centuries later, would form the foundations for the governance of contemporary western civilisation, with countries in other parts of the world also utilising similar governance systems.

The rise of Phillip II of Macedonia signalled the end of the independent Greek city states. It may be said that Phillip introduced civilisation into a kingdom in a poor part of northern Greece, which the other Greek states considered backward.[107] However, when Phillip seized power and became King in 359 BC after the death of his two older brothers, he set about modernising Macedonia.[108] In particular, Phillip set about modernising the Macedonian army using his experience gained from the Theban general Epaminondas, while he had been exile

[100] Ibid.

[101] Ibid.

[102] Osborne, R. (1996), Greece in the Making, 1200–479 BC, Routledge, New York.

[103] Ibid.

[104] Gaeta, L. (2004) Athenian democracy and the political foundation of space, Planning Theory & amp; Practice, 5:4, pp. 471–483.

[105] Ibid.

[106] Ibid.

[107] Cawthorne, N. (2004), Alexander the Great, Haus Publishing, London.

[108] Ibid.

in Thebes.[109] The Macedonian army was organised into an elite cavalry, behind which was a powerful phalanx of six brigades of infantry.[110] Each brigade contained 1500 men, with each man carrying a pike which was much longer than that used by the military of other Greek states.[111] This innovation may have given the Macedonians an advantage in battle, by disabling oncoming cavalry or infantry without Macedonian cavalry or men being in harm's way. The Macedonian infantry was also densely packed, such that injured men could be quickly replaced in the heat of battle, encouraging a continuous forward motion of soldiers. To support siege warfare, the Macedonian army would be followed by siege warfare engineers.[112]

The convening of the Amphictyonic Council, an ancient religious association of Greek tribes predating the rise of the Polis, in 339 BC signalled the start of the Fourth Sacred War against the Polis of Amphissa.[113] The meeting was convened because the Amphissans accused Athens of impiety. However, the Athenians in turn accused the Amphissan's of impiety by cultivating part of the sacred plain of Cirrha.[114] This was confirmed by the Council who appointed Phillip II, Alexander the Great's father, as the head of the Amphictyonic Council army, which was to be sent to punish the Amphissans.[115] Nevertheless, Athens and Thebes did not send representatives to the special Council meeting at which Phillip II was appointed commander of the Council's army.[116] This was for two reasons. Firstly, the Thebans did not want to wage war against the Amphissan's.[117] Secondly, the Athenians did not want to compromise their relationship with the Thebans by backing Phillip II, and this view

[109] Ibid.

[110] Ibid.

[111] Ibid.

[112] Ibid.

[113] Worthington, I. (2013), Demosthenes of Athens and the Fall of Classical Greece, Oxford University Press, Oxford.

[114] Ibid.

[115] Ibid.

[116] Ibid.

[117] Ibid.

was supported by the Athenian politician Demosthenes.[118] It was he who saw the threat which Phillip II posed to the independence of the Greek city states. For this reason, Demosthenes wanted to form an alliance with Thebes to defeat Phillip II.[119] Thebes was not on friendly terms with Macedonia. For example, the Thebans had seized the fortress of Nicaea from the Macedonian garrison placed there by Phillip in 344 BC.[120] But while Phillip was in control of the Amphictyonic army, he was free to do as he liked in Greece. Thus, Phillip seized the town of Elatea, and demanded that the Thebans return Nicaea to its rightful owners.[121] Furthermore, Phillip also demanded that the Thebans join his army and the Macedonians in a war to defeat the Athenians, in return for keeping the booty for the sacking of the city.[122] However, the Thebans stuck with the Athenians, but even their combined forces would not have been enough to stop Phillip and the Macedonians from overrunning the city states of Greece. The Athenians and the Thebans, therefore put out a call to the other Greek city states receiving support from Megara, Corinth, Achaea, Euboea, Acarnania, and some outlying island.[123] On the other hand, the Thessalians and the Phocians supported Phillip and his forces.[124] These opposing forces met at the Battle of Chaeronea in 338 BC, whereupon Phillip II and his forces emerged as the victors.[125] It was at the Battle of Chaeronea that the Greek city states lost their independence.[126] However, Athens suffered its final naval defeat at the hands of the Macedonians in 322 BC.[127] Phillip then established the League of

---

118 Ibid.

119 Ibid.

120 Ibid.

121 Ibid.

122 Ibid.

123 Ibid.

124 Ibid.

125 Worthington, I. (2013), Demosthenes of Athens and the Fall of Classical Greece, Oxford University Press, Oxford.

126 Ibid.

127 Davies, J. (1993), Democracy and Classical Greece, Harvard University Press, Cambridge, Massachusetts.

Corinth.[128] This was formed to facilitate cohesion in Greece by placing each of the former city states on an equal footing, whether they had previously been an ally or an enemy to Phillip and the Macedonians.[129] The League of Corinth was also allow all the resources, manpower, and weapons manufacturing, to be focused on Phillip's military objective of challenging, containing, and eventually defeating the Persians. Furthermore, Phillip's strategy of making formerly weaker states stronger and formerly stronger states weaker had the effect of destroying the former system of political control that the independent former city states had enjoyed.[130] However, this strategy did not prevent a revolt in Thebes in 335 BC or a war with Sparta in 331 BC.[131] But it may have prevented the spreading of revolts and military action against the Macedonians from one former city state to others.[132] The impact of Phillip and the Macedonians on Greece was to unite it physically.[133] However, Greece remained politically disunited simply because local nationalism at the level of the city state would not converge with nationalism at a national level.[134] Nevertheless, the formation of the League of Corinth signalled the end of formal military operations by Phillip and the Macedonians in Greece.[135] They would not return there until after the assassination of Phillip, when his son Alexander and his forces would return to quell discontent. The focus of Macedonian military forces would turn to Anatolia, and then, under Alexander, to a far wider area stretching from Europe to the frontiers of India and beyond.[136]

[128] Worthington, I. (2013), Demosthenes of Athens and the Fall of Classical Greece, Oxford University Press, Oxford.

[129] Roebuck, C. (1948), The Settlements of Phillip II with the Greek States in 338 BC, Classical Philology, Vol. 43, No. 2, pp. 73–92.

[130] Ibid.

[131] Ibid.

[132] Ibid.

[133] Ibid.

[134] Ibid

[135] Ellis, J. (1976), Phillip II and Macedonian Imperialism, Princeton University Press, Princeton, New Jersey.

[136] Ibid.

Hellenistic Greece (323 BC to 31 BC)

The term 'Hellenistic' denotes the period of ancient Greek history that encompasses the death of Alexander the Great in 323 BC and the death of Cleopatra VII of Egypt in 30 BC.[137] Within two decades of Alexander's death, his generals carved up the empire he had created between themselves. Macedonia and Greece were taken by Antigonus (382 BC–301 BC) and his son Demetrius (336 BC to 283 BC).[138] Syria and the old Persian kingdom, whose frontiers lay in western Asia, were taken by Seleucus (358 BC to 281 BC), and Ptolemy (367 BC to 282 BC) took Egypt.[139] However, it took several decades after Alexander's death for the boundaries of the new kingdoms to become stable.[140] The Seleucids Empire, founded by Seleucus, was by far the biggest in terms of the size of territory, stretching from Syria as far as Afghanistan.[141] However, it continued to lose territory both in the west and in the east.[142] Moreover, the new rulers were not descended from royal bloodlines, and so the legitimacy for their Kingships were derived solely from their military experience and positions under Alexander.[143] The new political economy of Greece and the near east was a combination of Macedonian and near-eastern traditions, a combination of exogenous rulers taking over the rule of peoples with endogenous religious, cultural, social, economic, political, and military traditions.[144] The new political economy of governance emerged during the conquests of Alexander when he made Macedonian kingship the centre of governance in the lands he acquired.[145]

---

[137] Martin, T. (2013), Ancient Greece: From Prehistoric to Hellenistic Times, Yale University Press, New Haven.

[138] Ibid.

[139] Ibid.

[140] Ibid.

[141] Price, S. (2001), The History of the Hellenistic Period, IN The Oxford History of Greece and the Hellenistic World, Boardman, J., Griffin, J., and Murray, O. (Eds), Oxford University Press, Oxford.

[142] Ibid.

[143] Ibid.

[144] Martin, T. (2013), Ancient Greece: From Prehistoric to Hellenistic Times, Yale University Press, New Haven.

[145] Price, S. (2001), The History of the Hellenistic Period, IN The Oxford History of Greece and the Hellenistic World, Boardman, J., Griffin, J., and Murray, O. (Eds), Oxford University Press, Oxford.

The Hellenistic Age encompasses a time in which Greek traditions fluxed with the indigenous traditions of the eastern Mediterranean to form a multicultural social and cultural life.[146] This was facilitated by the Hellenistic kings encouraging Greeks to move to their kingdoms, introducing new ideas and customs into non-Greek societies.[147] This diffusion of ideas from Greece to the new Hellenistic kingdoms had its biggest impact in the densely populated regions of Egypt and Southwest Asia.[148] One of the innovative ideas that was transferred from Greece to the new Hellenistic kingdoms was that of the institution of the Gymnasion.[149] This was the ancient Greek equivalent of the contemporary 'school' in which both physical and intellectual achievements are supported.[150] Mental and physical fitness allowed for the dynamism of thought, facilitating innovation in a diverse range of areas. This included the law, in which institutional changes occurred during the second and third centuries BC, such that there was greater inter-city co-operation in the dispensation of justice.[151] In this case, judges from other cities could sit in on the judgement panels and tribunals of other cities. Inter-state co-operation in the dispensation of justice became more integrated during the Hellenistic Age, making it more difficult for foreigners to evade justice.[152] However, the Greeks also established new cities in the Hellenistic kingdoms, which also facilitated the Greek influence on other cultures.[153] On the other hand, this process had already begun under Alexander when his armies conquered other civilisations—this also facilitated the diffusion of the Greek language, lifestyles as well as Greek institutions, such as the theatre and the Gymnasion.[154] Other

---

[146] Martin, T. (2013), Ancient Greece: From Prehistoric to Hellenistic Times, Yale University Press, New Haven.

[147] Ibid.

[148] Ibid.

[149] Price, S. (2001), The History of the Hellenistic Period, IN The Oxford History of Greece and the Hellenistic World, Boardman, J., Griffin, J., and Murray, O. (Eds), Oxford University Press, Oxford.

[150] Ibid.

[151] Chamoux, F. (2002), Hellenistic Civilisation, Blackwell Publishing, London.

[152] Chamoux, F. (2002), Hellenistic Civilisation, Blackwell Publishing, London.

[153] Martin, T. (2013), Ancient Greece: From Prehistoric to Hellenistic Times, Yale University Press, New Haven.

[154] Thonemann, P. (2018), The Hellenistic Age: A Very Short Introduction, Oxford University Press, Oxford.

innovations in the Hellenistic Age were made in the arts, in architecture, and in warfare with the use of elephants in battles between two opposing armies.[155] Nevertheless, knowledge of the use of war elephants was acquired by the Macedonians and the Greeks during Alexander's incursions into the Hindu Kush kingdoms of ancient India.[156] In literature, Polybius (200 BC to 118 BC) innovated in the way he wrote, choosing to write only on the history of actions at a turning point in world history through his 'Histories of Rome'.[157] The strong link between political hegemony and cultural influence is evidenced centuries later by the European nations' influence on their colonies throughout the world.[158] A prime example would be the impact of the British Empire on indigenous societies from the seventeenth century AD onwards.

Despite their apparent stability in a cosmopolitan society, the Hellenistic kingdoms were losing ground to another rising power in the Mediterranean, the power of Rome. It was the poor decisions taken by the Macedonian kings, from a diplomatic as well as from a military perspective, starting in the third century BC, which encouraged the Romans to take an interest in Greece.[159] This led to the increasing dominance of Rome in Greek affairs. This led to the control of Macedonia by Rome by the middle of the second century BC when, in 168 BC, Rome conquered Macedonia[160] and Greece became a Roman province in 146 BC.[161] After

[155] Chamoux, F. (2002), Hellenistic Civilisation, Blackwell Publishing, London.

[156] Ibid.

[157] Godin, B., and Lucier, P. (2012), Innovation and Conceptual Innovation in Ancient Greece, INRS, Chaire Fernand Dumont Sur La Culture, Project on the Intellectual History of Innovation, Working Paper No.12.

[158] Price, S. (2001), The History of the Hellenistic Period, IN The Oxford History of Greece and the Hellenistic World, Boardman, J., Griffin, J., and Murray, O. (Eds), Oxford University Press, Oxford.

[159] Martin, T. (2013), Ancient Greece: From Prehistoric to Hellenistic Times, Yale University Press, New Haven.

[160] Price, S. (2001), The History of the Hellenistic Period, IN The Oxford History of Greece and the Hellenistic World, Boardman, J., Griffin, J., and Murray, O. (Eds), Oxford University Press, Oxford.

[161] Smith, W., and Greene, G. (1861), A History of Greece from the Earliest Times to the Roman Conquest, Harper Brothers Publishers, New York.

this time, Roman and Greek history became intertwined.[162] The end of the Hellenistic kingdoms was met by the rise of the Roman Empire under its first Emperor, Augustus (31 BC to 14 AD), and the demise of the Roman Republic.[163]

References

Andrewes, A. (1956), The Greek Tyrants, Hutchinson's University Library, London.

Andrewes, A. (1977), Kleisthenes' Reform Bill, CQ 27, pp. 246–7

Blok, J. (2018), Retracing Steps: Finding Ways into Archaic Greek Citizenship, IN Defining Citizenship in Archaic Greece, Duplouy, A., and Brock, R. (Eds), Oxford University Press, Oxford.

Bryant, J. (1996), Moral Codes and Social Structure in Ancient Greece: A Sociology of Greek Ethics from Homer to the Epicureans and Stoics, State University of New York Press, Albany.

Carpenter, R. (1966), Discontinuity in Greek Civilization. Cambridge University Press, Cambridge, UK.

Castleden, R. (2005), Mycenaeans, Routledge, New York.

Cawthorne, N. (2004), Alexander the Great, Haus Publishing, London.

Chamoux, F. (2002), Hellenistic Civilisation, Blackwell Publishing, London.

Cook, R. M. (1958), Speculations on the origin of coinage, Historia 7, 257–62.

Davies, J. (1993), Democracy and Classical Greece, Harvard University Press, Cambridge, Massachusetts.

Desborough, V.R. (1964), The Last Mycenaeans and Their Successors. Oxford University Press, Oxford.

De Souza, P. (1998), Towards thalassocracy? Archaic Greek naval developments, IN Archaic Greece: New Approaches and New Evidence, Fisher, N., and Van Wees, H. (eds), Duckworth with The Classical Press of Wales, London.

[162] Martin, T. (2013), Ancient Greece: From Prehistoric to Hellenistic Times, Yale University Press, New Haven.

[163] Price, S. (2001), The History of the Hellenistic Period, IN The Oxford History of Greece and the Hellenistic World, Boardman, J., Griffin, J., and Murray, O. (Eds), Oxford University Press, Oxford.

Drake, B. (2012), The Influence of Climatic Change on the Late Bronze Age Collapse and the Greek Dark Ages, Journal of Archaeological Science, 39, pp. 1862–1870.

Ehrenberg, V. (1937), When did the polis rise? JHS 57, 147–59

Ellis, J. (1976), Phillip II and Macedonian Imperialism, Princeton University Press, Princeton, New Jersey.

Fleck, R., and Hanssen, F. (2013), How Tyranny Paved the Way to Democracy: The Democratic Transition in Ancient Greece, The Journal of Law and Economics, Vol. 56, No. 2, pp. 389–416.

Fortson IV, B.W. (2004), Indo-European Language and Culture, an Introduction, Blackwell Publishing, Malden, MA.

Gaeta, L. (2004) Athenian democracy and the political foundation of space, Planning Theory & amp; Practice, 5:4, pp. 471–483.

Gagarin, M. (2003), Letters of the Law: Written Texts in Archaic Greek Law, IN: Written Texts and the Rise of Literate Culture in Ancient Greece, Yunis, H. (Eds), Cambridge University Press, Cambridge.

Godin, B., and Lucier, P. (2012), Innovation and Conceptual Innovation in Ancient Greece, INRS, Chaire Fernand Dumont Sur La Culture, Project on the Intellectual History of Innovation, Working Paper No. 12.

Grant, M. (1987), The Rise of the Greeks. New York: Charles Scribner.

Grote, G. (1856), History of Greece, Murray.

Hammond, B. (1895), Political Institutions and the Greeks, Cambridge University Press, Cambridge.

Holkeskamp, K. (1992), Written Law in Archaic Greece, Proceedings of the Cambridge Philological Society, 38, pp. 87–117.

Janse, M. (2002), Aspects of Bilingualism in the History of the Greek Language, IN: Bilingualism in Ancient Society: Language Contact and the Written Text, Adams, J., Janse, M., and Swain, S. (Eds), Oxford University Press, Oxford.

Jeffrey, L. (1976), Archaic Greece: The City-States c. 700–500 B.C, St. Martin's Press, New York.

Kagan, D. (1982), The dates of the earliest coins, Am.J. Archaeol. 86, 52–73.

Kraay, C. (1976), Archaic and Classical Greek coins. London.

Kurke, L. (1992), The Politics of Archaic Greece, Classical Antiquity, Vol. 11, No. 1, pp. 91–120.

Martin, T. (2013), Ancient Greece – From Prehistoric to Hellenistic Times, Yale University Press, New Haven.

Mazar, A. (1990), Archaeology of the Land of the Bible 10,000– 86 BCE. Doubleday, New York, NY.

Morris, I. (1986), Gift and Commodity in Archaic Greece, Man, New Series, Vol. 21, No. 1, pp. 1–17.

Morris, I. (1990), The Early Polis as City and State, IN City and Country in the Ancient World, Rich, J., and Wallace-Hadrill, A. (Eds), Routledge, London.

Morris, I. (2004), Economic Growth in Ancient Greece, Journal of Institutional and Theoretical Economics, Vol. 160, No. 4, pp. 709–742.

Osborne, R. (1996), Greece in the Making, 1200–479 BC, Routledge, New York.

Palaima, T.G. (2010), Linear B. In: Cline, E. (Ed.), The Oxford Handbook of the Bronze Age Aegean. Oxford University Press, Oxford, pp. 356–372.

Parker, V. (2007), Tyrants and Lawgivers, IN The Cambridge Companion to Archaic Greece, Shapiro, H. (Ed), Cambridge University Press, Cambridge.

Percy, W. (1996), Pederasty and Pedagogy in Archaic Greece, University of Illinois Press, Chicago.

Price, S. (2001), The History of the Hellenistic Period, IN The Oxford History of Greece and the Hellenistic World, Boardman, J., Griffin, J., and Murray, O. (Eds), Oxford University Press, Oxford.

Robinson, E. (1997), The First Democracies: Early Popular Government Outside Athens, Franz Steiner Verlag, Stuttgart.

Roebuck, C. (1948), The Settlements of Phillip II with the Greek States in 338 BC, Classical Philology, Vol. 43, No. 2, pp. 73–92.

Rose, P. (2012), Class in Archaic Greece, Cambridge University Press, Cambridge.

Salmon, J. (2003), Lopping off the heads? Tyranny, politics and the polls, IN The Development of the Polis in Archaic Greece, Mitchell, L., and Rhodes, P. (Eds), Routledge.

Sandars, N.K. (1987), The Sea Peoples: Warriors of the Ancient Mediterranean. Thames and Hudson, New York.

Sandys, J. (1912), Aristolelous, Athenaian, Politeia, The Lawbook Exchange Ltd, Union, New Jersey.

Sass, B. (2005), The Alphabet at the Turn of the Millennium: The West Semitic Alphabet CA. 1150–850 BCE, the Antiquity of the Arabian, Greek, and Phrygian Alphabets, Ravgon, Tel Aviv.

Saxonhouse, A. (1988), The Tyranny of Reason in the World of the Polis, American Political Science review, Vol. 82, No. 4.

Sealey, R. (1976), A History of the Greek City States, 700–338BC, University of California Press, Los Angeles.

Shepherd, W. (2012), Plataea 479 BC: The Most Glorious Victory Ever Seen, Osprey Publishing.

Smith, W., and Greene, G. (1861), A History of Greece from the Earliest Times to the Roman Conquest, Harper Brothers Publishers, New York.

Snodgrass, A. (1981), Archaic Greece, The Age of Experiment, University of California Press, Berkeley and Los Angeles.

Strauss, B. (2004), The Battle of Salamis: The Naval Encounter That Saved Greece and Western Civilisation, Simon & Schuster Paperbacks, New York.

Thonemann, P. (2018), The Hellenistic Age: A Very Short Introduction, Oxford University Press, Oxford.

Ulf, C. (2009), The World of Homer and Hesiod, IN A Companion to Archaic Greece, Raaflaub, K., and Van Wees, H. (Eds), Blackwell Publishing Ltd.

Walter, U. (2010), The Classical Age as a Historical Epoch, IN A Companion to the Classical Greek World, Kinzl, K. (Ed), Blackwell Publishing Ltd, Oxford.

Worthington, I. (2013), Demosthenes of Athens and the Fall of Classical Greece, Oxford University Press, Oxford.

6

The Roman Empire (443 BC to 395 AD)

Early Rome and the Republic

Not much has been written about the beginnings of Rome, because most historical records have not survived. Much of early Roman history has been pieced together from a combination of myth, records of scholars who were active many centuries after the collapse of the Roman Empire, and from the fragments of texts of scholars who were contemporary to Rome. The first of these include Titus Livius (59 BC to 17 AD) and Dionysius of Halicarnassus, his Greek contemporary. Both ancient scholars provide parallel accounts of early Rome until 443 BC, but after this year, only Titus Livius continued to do so.[1] On the other hand, the works of Livius and Dionysius is augmented by the works of Cicero (106 BC to 43 BC) such as 'On the State', which provides a narrative on early Roman history, despite the fact that the text did not survive in its entirety.[2] Nevertheless, Plutarch (46 AD–120 AD) draws upon the works of Livius and Dionysius for his own narratives, but does make his own contribution

[1] Cornell, T. (1995), The Beginnings of Rome: Italy and Rome from the Bronze Age to the Punic Wars (c. 1000–264 BC), Routledge, New York.

[2] Ibid.

S. Ramesh, *The Rise of Empires*, https://doi.org/10.1007/978-3-030-01608-1_6

to Rome's historical narrative in the Life of Pyrrhus, covering the period 293 to 264.[3] The works of Diodorus Siculus, especially books 11–20 survived, and provide insights into early Rome from 486 BC to 302 BC.[4] Furthermore, fragments of surviving texts by Cassius Dio, as well as a summary of it by the twelfth-century Byzantine monk Zonaras, provide an account of early Rome up to the Punic Wars.[5] Book 5 of the works of the scholar Strabo (63 BC to 21 AD) also provides an insight into the very early history of Rome.[6] Other ancient scholars, such as Polybius (210 BC to 131 BC) and Cornelius Tacitus (56 AD to 120 AD), also contributed to the study of the history of Rome, but in different times.[7] The work of Polybius tells of the rise of Rome up to the Punic Wars with Carthage. His work is considered important because it is 100 years older than are the other existing accounts of classical antiquity.[8] Moreover, Polybius is not only able to shed light on the nature of treaties between Rome and Carthage, but also for its early history, extending back to 450 BC.[9] Finally, the work of Cornelius Tacitus sheds light on the evolution and development of institutions in ancient Rome.[10]

There has been some debate in the literature as to the specific date at which Troy was destroyed by the Mycenaeans. However, it may be the case that Troy was destroyed at some point within the range of dates encompassing the thirteenth to the twelfth centuries BC.[11] After the destruction of Troy, the chief associate of Hector, Aeneas, fled with the other Trojans to the west coast of Italy.[12] The story of how Aeneas goes to the west coast of Italy and what he does there subsequently is told by the

[3] Ibid.
[4] Ibid.
[5] Ibid.
[6] Ibid.
[7] Ibid.
[8] Ibid.
[9] Ibid.
[10] Ibid.
[11] Cline, E. (2013), The Trojan War – A Very Short Introduction, Oxford University Press, Oxford.
[12] Duncan, M. (2016), The History of Rome – Volume 1: The Republic, Herodotus Press.

ancient scholar Vergil in his epic, the 'Aeneid'.[13] Once Aeneas landed in Italy, he was met there by King Latinus who, rather than fight Aeneas and the Trojans, offered his daughter Lavinia in marriage.[14] Aeneas accepted the offer, but the problem was that Lavinia had already been promised in marriage to the leader of the Rutuli tribe. As a result, the Rutuli tribe attacked the Trojans and the Latins coalition, who successfully repelled and defeated them. However, as King Latinus was killed, the control of the Trojan and Latins coalition passed to Aeneas.[15] The remnants of the Rutuli then joined forces with another tribe, the Etruscans, to attack the Latin and Trojan coalition, but were again defeated by Aeneas. He then established the Tiber River as the border between the Etruscans and the Latins.[16] Asconius, the son of Aeneas, then followed him to become king, establishing a settlement at Alba Longa. Numitor followed Asconius as king, but he had a brother, Amulius, who also wanted to be king. The result was that Numitor was ousted and his sons killed, and his daughter Rhea was forbidden from marrying or to having any personal relations.[17] However, Rhea was taken advantage of and, as a result, gave birth to twin sons, Romulus and Remus. As they would pose a threat to his kingship, Amulius ordered the infants to be drowned. However, they were left amongst the reeds on the banks of the Tiber by a shepherd who brought the twins up as his own.[18] When Romulus and Remus grew into young men, they engaged in fighting low brigands. As a result, Remus was captured and taken to the stronghold of Numitor, who had become a local lord. Numitor then determined that Romulus and Remus were his grandsons. And with this knowledge, he seized back the throne of the Latins. Eventually, Romulus and Remus founded a new settlement on the banks of the Tiber, at the location where they had been abandoned.[19] Rome

[13] Neel, J. (2017), Early Rome: Myth and Society: A Sourcebook, John Wiley & Sons Inc., Hoboken, NJ, USA.

[14] Duncan, M. (2016), The History of Rome – Volume 1: The Republic, Herodotus Press.

[15] Ibid.

[16] Ibid.

[17] Ibid.

[18] Ibid.

[19] Ibid.

coalesced from a series of smaller settlements along the banks of the River Tiber, of which two of the main ones were the settlements of the Latins and the Sabines.[20] As time went on, there arose a conflict between Romulus and Remus as to who should be king. Because of the conflict, Romulus killed his brother Remus and became king of the city that was to be named after him, Rome, in 753 BC.[21] As it turns out, both Romulus and, many centuries later, Genghis Khan would have killed their brothers to achieve supremacy in their respective tribes. Romulus killed his twin and Genghis Khan killed his half-brother.

Between the ninth and the sixth centuries BC, Rome was one of several powerful settlements in the central Italy of the time.[22] The period that encompasses Rome as a republic is from 509 BC to 49 BC.[23] It was during this time that Rome acquired its empire.[24] The last King of Rome, Tarquin the Proud, was expelled in 507 BC.[25] When the King was away, the behaviour of his sons led L. Tarquinius Collatinus and L. Junius Brutus, his relatives, to lead a revolt that overthrew the monarchy, and the establishment of a republic.[26] The people were then freely able to elect a chief magistrate.[27] This would have been the equivalent of the consul, the highest form of magistrate.[28] Annually, two consuls with equal power were elected, both initially being of the Patrician class.[29] However, in the

[20] Raaflaub, K. (2006), Between Myth and History: Rome's Rise from Village to Empire (the eighth century to 264), IN: A Companion to the Roman Republic, Rosenstein, N., and Morstein-Marx, R. (Eds), Blackwell Publishing, Oxford, USA.

[21] Duncan, M. (2016), The History of Rome – Volume 1: The Republic, Herodotus Press.

[22] Lomas, K. (2017), The Rise of Rome, 1000 BC to 264 BC, Profile Books Ltd., London.

[23] Flower, H. (2014), Preface, IN: The Cambridge Companion to The Roman Republic, Flower, H. (Eds), Cambridge University Press, New York.

[24] Ibid.

[25] Raaflaub, K. (2006), Between Myth and History: Rome's Rise from Village to Empire (the eighth century to 264), IN: A Companion to the Roman Republic, Rosenstein, N., and Morstein-Marx, R. (Eds), Blackwell Publishing, Oxford, USA.

[26] Mackay, C. (2004), Ancient Rome: A Military and Political History, Cambridge University Press, Cambridge.

[27] Wiseman. T. (1998), Roman Republic Year One, Greece & Rome, Vol. XLV, No. 1.

[28] McCarty, N. (2008), Rome – The Greatest Empire of the Ancient World, The Rosen Publishing Group, Inc., New York.

[29] Ibid.

fourth century BC, in order to ensure that the Plebeians were not unjustly treated, one of the two consuls had to be a Plebeian.[30] Tribunes were also magistrates with both political and military duties.[31] They were elected by the Comitia Tributa; and only those born Plebeians or elected by the people could become tribunes.[32] The tribunes had the power to arrest and fine people, to organise public meetings, as well as to veto decisions made by the consuls.[33] The military power of the tribunes lay with the Roman Army.[34] Following the period of adjustment after the end of kingship, the time during which Rome was a republic could be broken down into further specific time periods. For example, the period 494 BC to 450 BC, during which there was no written legal code—the Twelve Tables.[35] The period of the consular tribunes, 450 BC to 366 BC, and the period of the Patricians and the Plebeians, 366 BC to 300 BC,[36] and notably, the period of dictatorship by Caesar between 49 BC and 44 BC.[37] There is nothing to suggest that, within the time that Rome was a republic, all of central Italy's settlements would have been acquired by it so that it would be recognised as an empire by the second century BC.[38]

The Twelve Tables represented the first form of codified law, which also formed the basis of all subsequent Roman Law.[39] It arose because of the conflict between the Patricians and the Plebeians, and laid down rights and fixed penalties, so that the law could not be used as a means of oppression.[40] The Patricians were the ruling class of families from the earliest days of the birth of the republic, while the Plebeians represented

---

[30] McCarty, N. (2008), Rome – The Greatest Empire of the Ancient World, The Rosen Publishing Group, Inc., New York.

[31] Ibid.

[32] Ibid.

[33] Ibid.

[34] Ibid.

[35] Flower, H. (2014), Preface, IN The Cambridge Companion to The Roman Republic, Flower, H. (Eds), Cambridge University Press, New York.

[36] Ibid.

[37] Ibid.

[38] Lomas, K. (2017), The Rise of Rome, 1000 BC to 264 BC, Profile Books Ltd., London.

[39] Conant, E. (1928), The Laws of the Twelve Tables, 13, St Louis L.Rev.231.

[40] Ibid.

the oppressed general citizenry. The battle for equality between the Patricians and the Plebeians reached a climax in the mid-fifth century BC.[41] It was at around this time that the Twelve Tables came into existence. They represented the rights of the citizens and the penalties erring citizens would be subject to, inscribed on bronze or copper plates fastened to pillars in public spaces.[42] In themselves, the Twelve Tables did not represent an enactment of new laws, but rather, the codification of existing laws. Each of the Tables dealt with a different aspect of the law. For example, Table 3 addressed the rights of creditors,[43] Table 4 dealt with the legal aspects of paternal power,[44] Table 5 dealt with succession and guardianship.[45] Table 6 dealt with the legal aspects of ownership and possession.[46] In this case, with regards to the rights of possession, Table 6 stipulated that for land, this would be qualifiable after 2 years, but for all other property, 1 year.[47] On the other hand, Table 9 dealt with the aspects of Public Law.[48] While the plates depicting the Twelve Tables were destroyed by the invading Gauls in 390 BC, the legal codification embodied in the Twelve Tables were themselves embodied in the Code of Justinian.[49] This put together old imperial statutes in one collection, the Codex, but also consisted of two other parts, the 'Digest' and the 'Institutes'.[50] The writing of the Code of Justinian was ordered by the Emperor Justinian of the eastern Roman Empire in the sixth century AD.[51] And it was at this time that the unwritten legal code that all men should be free to navigate the sea and use its resources was finally

[41] Ibid.
[42] Ibid.
[43] Ibid.
[44] Ibid.
[45] Ibid.
[46] Ibid.
[47] Ibid.
[48] Ibid.
[49] Ibid.
[50] Shapiro, M. (1981), Courts: A Comparative and Political Analysis, The University of Chicago Press, Chicago.
[51] Ibid.

codified.[52] Following the Twelve Tables in the fifth century BC, the development of Roman Law centred upon two pivotal events in Roman history.[53] The first was the expansion of the frontiers of Roman control in the Mediterranean following the end of the second Carthaginian War (218 BC–200 BC). The second event was the death of the Emperor Alexander Severus (235 AD). These two historical events mask three different ages associated with three different law systems.[54] These being strictly Roman Law, the universal Roman Law and the Greek-Roman Law.[55] Strictly Roman Law applied in the first period during which Rome was a semi-closed rural society. On the other hand, universal Roman Law and Greek-Roman Law represented a shift towards a more progressive and open society, with the latter gaining ground under the Emperor Justinian 1 of the eastern empire.[56] As the legal changes progressed over several centuries, the political structures of Roman society shifted from those necessary to support a rural community to one which could support a complex central government. In this case, there was a shift away from familiae (family), gentes (group of family with a common ancestor), and tribus (original tribes of Rome, the Ramnes, Tities, and the Luceres) to a nuclear family and a centralised form of government.[57] As the political structures of Rome shifted away from clan structures, to one which was based on a nuclear family, there was progressively greater emphasis on the rights of the individual. This led to a certain degree of emancipation for women, who increasingly became free from the protection of guardians, which had restricted their freedom.[58]

At the heart of the military success of the Roman Republic may have been its constitution. Indeed, the scholar Polybius positively correlates

[52] Fenn, P. (1925), Justinian and the Freedom of the Sea, The American Journal of International Law, Vol. 19, No. 4, pp. 716–727.

[53] Biagini, E. (1994), Roman law and Political Control – from a Primitive Society to the Dawn of the Modern World, GeoJournal, 33.4, pp. 331–340.

[54] Ibid.

[55] Ibid.

[56] Ibid.

[57] Ibid.

[58] Ibid.

the strength of the constitution with military success.[59] This is because the constitution of the Roman Republic consisted of elements of the monarchy and democracy, as well as that of the patriarchy.[60] The combination of different governance structures had two consequences.[61] Firstly, it ensured that the freedom of the Roman citizens could be safe from the consequences of autocratic power. Secondly, freedom allowed the Roman Republic to project its own freedoms by shaping its own destiny. Freedoms allowed the best and the brightest of Roman citizens to enter public life.[62] Furthermore, freedom lends itself to a more dynamic society which is creative and innovative from many perspectives, including the political, social, technological, managerial, as well as the economic.[63] This contrasts with a society with less individual freedom, in which creativity and innovation would be constrained because of a lack of incentive to be so. However, while western literature denotes Rome as the starting point of innovation, the history of invention and innovation is far older in the context of the Indian subcontinent, for example, from where the concept of 'zero' emanates.[64]

The scholar-philosopher Plato, in his 'Republic', states that the objective of any city state's constitution should be military success.[65] However, the sixth king, Servius Tullius, made an institutional change to the military.[66] This was the adoption of the 'hoplite phalanx', which was replaced in the fifth and in the fourth centuries by the manipolar system, which

[59] Lintott, A. (1999), The Constitution of the Roman Republic, Oxford University Press, Oxford.

[60] Scott-Kilvert, I. (1979), The Rise of the Roman Empire, Polybius; translated by Ian Scott-Kilvert; selected with an Introduction by F.W. Walbank, Penguin Classics, Harmondsworth.

[61] Ibid.

[62] Buttle, N. (2001), Republican Constitutionalism: A Roman Ideal, The Journal of Political Philosophy, Vol. 9, No. 3, pp. 331–349.

[63] Biagini, E. (1994), Roman law and Political Control – from a Primitive Society to the Dawn of the Modern World, GeoJournal, 33.4, pp. 331–340.

[64] Sahay, A. (2005), Review Essay: Managing Innovation and Technology, The Journal of Business Perspectives, Vol. 9, No. 4.

[65] Lintott, A. (1999), The Constitution of the Roman Republic, Oxford University Press, Oxford.

[66] Mackay, C. (2004), Ancient Rome: A Military and Political History, Cambridge University Press, Cambridge.

was more flexible.[67] This involved heavily armed infantry soldiers equipped with round shields moving towards the enemy in a closely ranked formation. The phalanx was composed of groups of 96 men, to a depth of 8 men and a width of 12 men.[68] The strategic advantage of the phalanx was that the frontal assault could continue even with the loss of men on the frontline, as they would be replaced by the second man in the line stepping forth.[69] To assist in the adoption of the hoplite phalanx, the organisation of the citizen assembly was adjusted by additional infantry units for older citizens.[70] Furthermore, the citizen body was adjusted into units distinguished by tribe and district.[71] In the context of governance, decision making in early Rome was based around the King, the experienced elders (Patricians) and an assembly of land-owning families.[72] The coalescence of settlements had led to the formation of Rome and, by the seventh century BC, it had become a unified city.[73] New innovations included a paved forum area in the centre, encompassing an assembly area (comitium), a meeting place for the Senate (curia), the seat of the King (Rex), as well as shrines and sanctuaries.[74] The expansion of public spaces continued with the development of the 'cattle market', more sanctuaries throughout the city as well as on the Capitol, upon which the construction of the temple of Capitoline Triad was begun in the latter part of the sixth century BC.[75]

---

[67] Raaflaub, K. (2006), Between Myth and History: Rome's Rise from Village to Empire (the eighth century to 264), IN: A Companion to the Roman Republic, Rosenstein, N., and Morstein-Marx, R. (Eds), Blackwell Publishing, Oxford, USA.

[68] Keppie, L. (1998), The Making of the Roman Army: From Republic to Empire, University of Oklahoma Press, Norman.

[69] Ibid.

[70] Raaflaub, K. (2006), Between Myth and History: Rome's Rise from Village to Empire (the eighth century to 264), IN: A Companion to the Roman Republic, Rosenstein, N., and Morstein-Marx, R. (Eds), Blackwell Publishing, Oxford, USA.

[71] Ibid.

[72] Ibid.

[73] Ibid.

[74] Ibid.

[75] Ibid.

The year 264 BC would be a good starting point for the analysis of the evolution of Rome as a powerful city state in the Mediterranean, which would end the dominance of other powers in the region such as Carthage and the Greeks. It was in the year 264 BC that Rome sent an army abroad, for the first time, to confront the power of Carthage. This first war between Rome and Carthage heralded the rise of Rome and the eventual fall of Carthage and other powers in the Mediterranean region.[76] Between 400 BC and 300 BC, Rome controlled the majority of mainland Italy with control of the whole of the Mediterranean region following between 200 BC and 100 BC.[77] Much of the expansion of the republic occurred incrementally, whereby neighbouring cities, towns, and tribes were absorbed into Rome's orbit of control either through military superiority, the offer of military protection, or Roman citizenship. This was an individual status rather than as an individual capacity.[78] In this case, Roman citizens were not expected to take an active part in the democratic process. However, they were not expected to be passive either. For Roman citizens who did participate in the democratic process, this participation took two forms.[79] Firstly, they could take part in the decision making of the assemblies. Secondly, take part in the elections for the magistrates and the consuls.

The incremental growth of the boundaries of the Roman Republic, and the fact that it had a standing army, meant that, in contrast to Qin dynasty China, Rome did not need to maintain a sophisticated and complex administrative structure to mobilise resources for military conflict.[80]

---

[76] Cornell, T. (1995), The Beginnings of Rome: Italy and Rome from the Bronze Age to the Punic Wars (c. 1000–264 BC), Routledge, New York.

[77] Rosenstein, N. (2009), War, State Formation, and the Evolution of Military Institutions in Ancient China and Rome, IN Rome and China: Comparative Perspectives on Ancient World Empires, Scheidel, W. (Ed), Oxford University Press, Oxford.

[78] Buttle, N. (2001), Republican Constitutionalism: A Roman Ideal, The Journal of Political Philosophy, Vol. 9, No. 3, pp. 331–349.

[79] Nicolet, C. (1980), The World of the Citizen in Republican Rome, University of California Press, Berkeley.

[80] Rosenstein, N. (2009), War, State Formation, and the Evolution of Military Institutions in Ancient China and Rome, IN Rome and China: Comparative Perspectives on Ancient World Empires, Scheidel, W. (Ed), Oxford University Press, Oxford.

On the other hand, in the case of Qin dynasty China and the warring states period, in particular, the state was continually threatened by foes, and there was a need to ensure that the state could mobilise its maximum resources in a very short period of time in order to combat imminent military threats.[81] As a result, in Qin dynasty China, there was a need to maintain a sophisticated and permanent administrative structure to maximise the mobilisation of the state's resources at short notice. However, in the case of the Roman Republic, public affairs were controlled by the aristocrats of the Senate, from whose body were drawn the magistrates.[82] This was an electable office, potentially dependent on the votes of the eligible population, but elections tended to be far from democratic due to institutional rigidities.[83] Nevertheless, eligible voters and those due to pay direct taxes were identified by means of a census, but its accuracy depended on the willingness of citizens to take part.[84] Direct taxes were abolished in 167 BC by the Senate, although the state recouped revenues lost through a set of indirect taxes.[85] While the Senate lacked formal authority, due to the social status of its members, its informal authority on Roman society was strong.

The Fall of the Roman Republic (32 BC to 30 BC)

The Roman Empire would extend from the west to the east. The western provinces could be divided into two zones, depending upon their distance from the zone and the nature of geographical barriers.[86] The first zone was the Mediterranean zone, in which regions were separated from

[81] Ibid.
[82] Ibid.
[83] Ibid.
[84] Ibid.
[85] Ibid.
[86] Leveau, P. (2007), The Western Provinces, IN: The Cambridge Economic History of the Greco-Roman World, Scheidel, W., Morris, I., Saller, R. (Eds), Cambridge University Press, Cambridge.

Rome by the sea.[87] The second zone was the oceanic and the continental zone, which was separated from Rome by the Alps.[88] This zone comprised the linguistic zones associated with the Germanic tribes, the Celts, the Iberians, and the Libyans, although the language of administration was Latin—the language of Rome.[89] A frontier zone existed between the Celts in Gaul and the Germanic tribes.[90] The latter proved to be difficult for the Roman Empire.[91] On the other hand, the eastern part of the Roman Empire, comprising the eastern Mediterranean, consisted of areas including modern Greece, the former Yugoslavia, the Republic of Macedonia, modern Turkey, Syria, Judea, and Arabia.[92]

When Julius Caesar crossed the Rubicon River in 49 BC, at the end of his tenure as the Roman Governor of a region stretching from Gaul to Illyricum, with the 13th Legion, it marked the beginning of the end of the Roman Republic and the rise of autocracy and Imperial Rome. Caesar had been appointed as Governor by the Senate of the Roman Republic. However, once his tenure as Governor had been completed, the Senate instructed Caesar not to march his troops back to Rome. But Caesar did not obey the Senate's instructions, and they saw this as treason. As a result, once Caesar and his troops had crossed the Rubicon River, a boundary between Rome and its provinces, it signalled the beginning of a twenty-year civil war and the end of the Roman Republic.[93] The reason why the republic came to an end was the ineffectiveness of traditions and institutions—the use of force in the place of laws and the constitution.[94] Chaotic governance substituted freedom for autocracy. The starting point

[87] Ibid.

[88] Ibid.

[89] Ibid.

[90] Cherry, D. (2007), The Frontier Zones, IN: The Cambridge Economic History of the Greco-Roman World, Scheidel, W., a Morris, I., Saller, R. (Eds), Cambridge University Press, Cambridge.

[91] Ibid.

[92] Alcock, S. (2007), The Eastern Mediterranean, IN: The Cambridge Economic History of the Greco-Roman World, Scheidel, W., Morris, I., Saller, R. (Eds), Cambridge University Press, Cambridge.

[93] Gruen, E. (1974), The Last generation of the Roman Republic, University of California Press, Berkeley.

[94] Ibid.

for an analysis of the end of the Roman Republic could start with the rise of Sulla the Dictator.[95] Sulla began his dictatorship of the Roman Republic in 82 BC following the end of the Social War between 88 BC and 82 BC.[96] Historically, the strength of both political practices and the institutions of the Roman Republic were built around the ideal of the freedom of the Roman people.[97] Moreover, the late republic political dialogue emphasised the ideal of the free Roman citizen living in a free state.[98] This ideal is very much prevalent in the present-day United States. As well as the basis upon which the Republic was founded, departing the tyranny of the monarchy of Great Britain in the latter part of the eighteenth century AD. The ideal of the freedom of the Roman citizen was the bedrock of the Roman Republic, and the source of its political and ideological legitimacy.[99] Furthermore, the ideal of freedom ensured that Roman citizens had access to at least a basic level of civil rights, such as the right of appeal against sentences determined by magistrates.[100] As well as the right to elect their own leaders, and the right to assistance from the tribunician.[101] The latter was chosen from the Plebeians, and accorded the right of veto against decisions made by Patrician magistrates and the Senate. The civil rights of Roman citizens, even at a basic level, enabled them to pursue their own goals, and so, protect the ideal of freedom.[102] Moreover, the civil rights of the ordinary citizen were a means by which individual goals could be achieved by overcoming the obstructive behaviour of the ruling Patrician class of families.[103] Roman citizens could also own property.[104]

---

[95] Ibid.

[96] Vervaet, F. (2004), The 'Lex Valeria' and Sulla's Empowerment as Dictator (82–79 BCE), Cahiers du Centre Gustave Glotz, Vol. 15, pp. 37–84.

[97] Mouritsen, H. (2004), Plebs and Politics in the Late Roman Republic, Cambridge University Press, Cambridge.

[98] Ibid.

[99] Ibid.

[100] Ibid.

[101] Ibid.

[102] Ibid.

[103] Ibid.

[104] Arena, V. (2012), Libertas and the Practice of Politics in the Late Roman Republic, Cambridge University Press, Cambridge.

In the years before Sulla's ascendancy, many factors may have played a part in the decay of republican institutions. Firstly, from a theoretical perspective power lay in the hands of the people in the assembly.[105] The latter was responsible for the passing of all laws as well as for the election of the magistrates.[106] Two consuls represented the leadership of the magistrates; and during their tenure assumed a range of uncontested powers.[107] In contrast to the assembly, the Senate was supposed to be a docile legislative body. The composition of the Senate encompassed those who had previously served as magistrates. It was the knowledge that they had accumulated in that capacity which gave them membership of the Senate whose sole purpose was consultative.[108] In this case, serving magistrates would call upon the members of the Senate for advice as the need arose. However, at the time Sulla entered the scene, the Senate had assumed upon itself a directorial government role for which there had been no legislation by the people's assembly.[109] The usurped dominance of the Senate arose from the assumed superior knowledge of the ex-magistrates who made up its body, which allowed its opinions to substitute for the laws of the republic.[110] Lastly, the final factor that contributed to the instability and the decay of the institutions of the Roman Republic related to the political geography of the Italian mainland itself. This constituted a federation of states with the power of the Roman Republic at its heart.[111] However, it was very much an unequal confederation because only Roman citizens enjoyed the benefits of citizenship, a perk denied to non-Romans. This infuriated other states on the Italian mainland that provided a significant contribution of men to the Roman Army. Thus, the Social War, 91 BC to 89 BC, may have resulted not just because of a desire to be treated equally as citizens of Rome, but perhaps a quest for

[105] Keaveney, S. (2005), The Last Republican: Sulla, Routledge, New York.

[106] Ibid.

[107] Ibid.

[108] Ibid.

[109] Ibid.

[110] Ibid.

[111] Ibid.

total independence from Roman control.[112] However, after the first few months of the war, the people of Rome passed the first laws that would allow other peoples of the Italian mainland to share some of the rights of Roman citizens.[113] Nevertheless, the implementation of the rights won during the Social War took place over the next two to three centuries.[114] For example, Roman citizenship did not become widely available until 212 AD, although common people were not well treated.[115] A formal thought process or logic, which had been distinguished from language by the Greeks, therefore proved to be an effective tool for the unification of the Italian mainland and the Roman Republic.[116] However, in the context of formal thought, the Romans of the republic failed to make a distinction between science and philosophy.[117] But Aristotle stated that, while philosophy consisted of 'confused aggregates', science involved the determination of the elements and principles of those 'confused aggregates' through formal analysis.[118]

The Zenith of the Roman Empire (117 AD)

The end of the Roman Republic was followed by the rise of the early Roman empire in 27 BC.[119] This was based on a political structure known as a principate, which was in effect monarchic, developed under

[112] Ibid.

[113] Millar, F. (1986), Politics, Persuasion and the People before the Social War, Journal of Roman Studies, lxxvi, pp. 1–11.

[114] Nicolet, C. (1980), The World of the Citizen in Republican Rome, University of California Press, Berkeley.

[115] Biagini, E. (1994), Roman law and Political Control – from a Primitive Society to the Dawn of the Modern World, GeoJournal, 33.4, pp. 331–340.

[116] Moatti, C. (2015), The Birth of Critical Thinking in Republican Rome, Cambridge University Press, Cambridge.

[117] Ibid.

[118] Frank, P. (2004), Philosophy of Science: The Link Between Science and Philosophy, Dover Publications Inc., New York.

[119] Temin, P. (2006), The Economy of the Early Roman Empire, Journal of Economic Perspectives, Vol. 20, No. 1, pp. 133–151.

Augustus.[120] In the early Roman Empire, the Italian peninsula was only 30% urbanised.[121] Slaves were also a feature of the Roman Empire. However, slaves could become free if they worked hard, developed specialist skills, or bought their own freedom.[122] Nevertheless, despite the fact that slaves could keep any money they earned and accumulated, they still could own property, which was a privilege for only Roman citizens.[123] On the other hand, slaves who became free kept connections with as well as the names of their former owners.[124] This had advantages for slave-owning families because productive slaves would be readily able to increase the incomes of their former masters.[125] Furthermore, former slaves were able to marry Roman citizens, and their children and grandchildren arising from such a formal union were readily accepted into Roman society.[126] The allure of freedom incentivised slaves to work hard, progress, and gain their freedom.[127] From an economic perspective, a freed slave was just as attractive to commercial agents as a freeman because such freed slaves were able to act as commercial agents as well as transact business as merchants.[128] In the early Roman Empire, slavery was a formally enforceable labour contract, the benefits of which attracted poor people as a means to better their lives.[129] The labour market in the early Roman Empire had slaves being used alongside waged freemen.[130] In this case, there can be no doubt that the labour market in the Roman Empire was a unified one with the level of wage dispersion being comparable to that of pre-industrial Europe.[131]

---

[120] Ibid.

[121] Hopkins, K. (1978), Conquerors and Slaves, Cambridge University Press, Cambridge.

[122] Temin, P. (2006), The Economy of the Early Roman Empire, Journal of Economic Perspectives, Vol. 20, No. 1, pp. 133–151.

[123] Ibid.

[124] Ibid.

[125] Ibid.

[126] Ibid.

[127] Ibid.

[128] Ibid.

[129] Ibid.

[130] Ibid.

[131] Temin, P. (2004), The Labour Market of the Early Roman Empire, Journal of Interdisciplinary History, XXXIV:4, pp. 513–538.

In an economic context, Romans also loaned money to each other, with the loans being used for either production or consumption.[132] In this case, loans were also made to finance trade, with Roman merchants acquiring loans on the condition of the safe return of trading ships from other parts of the world.[133] However, the interest rate charged for this type of loan was particularly high.[134] Loans were often obtained from banks in places such as Delos in Greece, where banks had been operating before Rome, as well as after the Roman conquest.[135] The Romans patronised Delos because of its economic importance to the empire, allowing it to develop as a major free port in the Mediterranean while other commercial urban centres such as Carthage and Corinth were destroyed by military force.[136] Nevertheless, not many private banks acted as financial intermediaries in the exchange of pooled funds between savers and borrowers.[137] But it is apparent that financial markets in the world of the Roman Empire were connected, with losses in Asia contributing to losses at home.[138] This interconnectedness of Roman financial markets could have arisen from the fact that loans in one part of the Roman world became deposits in another part of the Roman world. And at a Macroeconomic level, the Roman Empire was not run based on the imperial government borrowing money, as is now done in modern economies, but on the basis of accumulated tax revenues.[139] This may explain why the Roman Empire neither had a national debt nor a central bank, which would have been responsible for raising loans to finance state

[132] Temin, P. (2006), The Economy of the Early Roman Empire, Journal of Economic Perspectives, Vol. 20, No. 1, pp. 133–151.

[133] Ibid.

[134] Ibid.

[135] Temin, P. (2006), The Economy of the Early Roman Empire, Journal of Economic Perspectives, Vol. 20, No. 1, pp. 133–151.

[136] Glynn, D. (2002), The Development of Greek and Roman Money, 600 BC–610 AD, History of Money, University of Wales Press.

[137] Temin, P. (2004), Financial Intermediation in the Early Roman Empire, The Journal of Economic History, Vol. 64, No. 3, pp. 705–733.

[138] Temin, P. (2006), The Economy of the Early Roman Empire, Journal of Economic Perspectives, Vol. 20, No. 1, pp. 133–151.

[139] Ibid.

expenditure in the advent of shortfalls in tax revenues.[140] The legal, financial, and governance institutions of the early Roman Empire allowed for stability and safety as more lands were conquered, and as the market for Roman traders expanded. At the time of the Roman Empire, information for merchants was at a premium. In the case of Roman grain merchants, who may have had to import large stocks of grain for Rome's needs, information asymmetry was overcome using economic and social institutions so that transaction costs and uncertainty would be reduced.[141] In the long run, security, trade, and tax collection facilitated economic growth.[142] Moreover, there is no doubt that the consumption and production encapsulated a market economy in the Roman Empire in which exchanges of goods and services were on the basis of payment in coins.[143] Furthermore, there is no doubt that the Roman Empire was a monetised economy, with people using a standard unit of value as payment for the exchange of goods and services throughout the lands of the empire.[144] The use of coinage developed in Lydia and Ionia and was inherited by the Roman Empire through its contacts with Greece.[145] However, it is indeterminate whether coinage was used in the Indian subcontinent, although it is highly likely that it was, given the extent of trade between east and west. This was especially true between the sixth century BC and the fourth century BC, when there was a great intermixing of cultures.[146] But it is known that there was contact between the Mauryans of north India and the Hellenistic kingdoms of the Ptolemies and the Seleucids

---

[140] Temin, P. (2004), Financial Intermediation in the Early Roman Empire, The Journal of Economic History, Vol. 64, No. 3, pp. 705–733.

[141] Kessler, D., and Temin, P. (2007), The organisation of the Grain Trade in the early Roman Empire, Economic History Review, 60, 2, pp. 313–332.

[142] Temin, P. (2006), The Economy of the Early Roman Empire, Journal of Economic Perspectives, Vol. 20, No. 1, pp. 133–151.

[143] Ibid.

[144] Kessler, D., and Temin, P. (2005), Money and Prices in the Early Roman Empire, Working Paper 05–11, MIT.

[145] Davies, G. (2002), The Development of Greek and Roman Money, 600 BC–610 AD, History of Money, University of Wales Press.

[146] Ibid.

during the third century BC is known.[147] Nevertheless, Roman maritime trade represented the most significant contact between the cultures of Asia and Europe.[148] And merchant trading ships from southern India even sailed along the Red Sea to the port of Alexandra.[149]

The Empire did not have a central banking system, but did have a mint to produce silver coinage, managed by the Roman Treasury, but under the direct control of the Emperor.[150] One of the functions of the Treasury was to maintain reserves and, in association with the reserves held by private banks such as the one at Delos, it was possible to control the level of inflation.[151] This unit of payment, the silver coinage produced by the Roman Mint, was represented by the Sesterii in the western empire and by the Drachmae in the eastern empire, both having a fixed exchange rate between them.[152] The result of this was that the market area encompassed by the Roman Empire had the characteristics of a common currency area.[153] However, the extent of this market economy was restricted by lack of technology and the lack of infrastructure and population size in the context of today's globalised market economy. But the use of a common currency and the imposition of political structure by the Romans in their empire allowed trade to expand and unify the price level between different regions of the empire, to some extent.[154] Nevertheless, the Roman Empire of the Mediterranean traded with other civilisations of the time, such as the Chola Dynasty in Southern India.

[147] Thapar, R. (1988), Epigraphic evidence and some Indo-Hellenistic contacts during the Mauryan period, in Maity, S.K., and Thakur, U. (eds.), *Indological Studies, Prof. D. C. Sircar Commemoration Volume*, New Delhi, pp. 15–19.

[148] Thapar, R. (1992), Black Gold: South Asia and the Roman Maritime Trade, South Asia: Journal of South Asian Studies, 15:2, pp. 1–27.

[149] Ibid.

[150] Davies, G. (2002), The Development of Greek and Roman Money, 600 BC–610 AD, History of Money, University of Wales Press.

[151] Ibid.

[152] Ibid.

[153] Ibid.

[154] Ibid.

Stability also improved throughout the Roman Empire once pirates had been cleared from the Mediterranean in 67 BC.[155] In contrast, while the Greek city states also promoted stability through efficient political conditions, the lands ruled by the Greek city states were much smaller in size.[156] The markets were, therefore, also much smaller. The early Roman Empire came to an end at the beginning of the third century AD.[157] This occurred for many reasons associated with many autocratic and brutal short-term emperors, army revolts, and destabilising and rampantly rising inflation in the second century AD. The banks that had financed much of imperial trade in Rome also disappeared during the third century AD.[158] Furthermore, the plague spread throughout the empire in the second century AD, having started at a military camp in the east.[159] Successive waves of the plague also spread throughout the empire for a twenty-five-year period, from around 250 AD.[160] The impact of the disease would have reduced the capacity of the military as well as a reduction in economic activity. However, the reign of Aurelian between 270 AD and 275 AD offered the empire some recovery from the crisis it had been facing. This was specifically true because of Aurelian's political, financial, and religious innovations which helped to bring Romans together. At a political level, Aurelian gave amnesties to those who had committed political offences against the Roman state as well as being opposed to it.[161] Furthermore, Aurelian gave amnesties to those in debt to the Roman state by having such financial records burned publicly in the forum.[162] Lastly, Aurelian established the Cult of Sol, and this became Aurelian's personal deity.[163]

[155] Temin, P. (2006), The Economy of the Early Roman Empire, Journal of Economic Perspectives, Vol. 20, No. 1, pp. 133–151.

[156] Ibid.

[157] Ibid.

[158] Ibid.

[159] Giardina, A. (2007), The Transition to Late Antiquity, IN: The Cambridge Economic History of the Greco-Roman World, Scheidel, W., Morris, I., Saller, R. (Eds), Cambridge University Press, Cambridge.

[160] Ibid.

[161] Dmitriev, S. (2004), Traditions and innovations in the Reign of Aurelian, Classical Quarterly, 54.2, pp. 568–578.

[162] Ibid.

[163] Ibid.

The later Roman Empire began in 200 AD, and its main characteristics were economic and political instability.[164] The decline of the Empire was matched by increasing centralisation of power in the government and greater government control over both social and economic activity.[165] The latter ranged from the control of prices to the control of agricultural production.[166] At the same time, private enterprise, innovation, and creativity became progressively static with no significant change.[167] Centralisation of power became the dominant political structure, and while the evolution of private enterprise progressed, the social structure in towns and cities were less open to innovation.[168] It would take many centuries after the demise of Rome for the individual values realised during its time to be coupled with innovation during the Industrial Revolution in late eighteenth-century England. This in turn would lead to rapid technological advancement over the subsequent 200 years, rising incomes, rising populations, the advancement of human rights, and a more connected world. The reason why innovation and individual freedoms did not lead to an Industrial Revolution in Roman times may have been because printing technology was not available. In this case, ideas could not be shared leading to a lack of innovation.[169] As a result, knowledge could not be readily accumulated and built upon over time. Nevertheless. Some interesting innovations did emerge from the Roman Empire.[170] The Romans were the first to develop and use concrete for construction. They also developed the arch for use in construction and a water pumping system that could be used to pump water for irrigation, bathing, and drinking simultaneously.[171] However, the problem was that the Roman

[164] Goodman, M. (1997), The Roman World, *44 BC–AD 180.* London, Routledge.

[165] Biagini, E. (1994), Roman law and Political Control – from a Primitive Society to the Dawn of the Modern World, GeoJournal, 33.4, pp. 331–340.

[166] Ibid.

[167] Ibid.

[168] Ibid.

[169] David, P. (1998), Common Agency Contracting and the Emergence of 'Open Science' Institutions. American Economic Review. 88:2, pp. 15–21.

[170] Temin, P. (2006), The Economy of the Early Roman Empire, Journal of Economic Perspectives, Vol. 20, No. 1, pp. 133–151.

[171] Ibid.

Empire used lead for the construction of water pipes.[172] Similarly, the Romans of the time also used lead for making cooking utensils.[173] But the problem with this was that it made them susceptible to lead poisoning.[174] Lead poisoning in early life will have disastrous effects on mental development[175] and, over time, this may have contributed to the decline of Roman civilisation.[176] The forensic analysis of the bones of ancient Romans supports this view because the bone concentrations of lead are significant. The Romans used water wheels for grounding grain and powering saw mills.[177] They also used hydraulic mining techniques which remained unsurpassed until the nineteenth century AD.[178] In the arts, the Romans also showed innovations, for example, by depicting allegorical figures, the symbols of war, and those of religious significance on sculptures.[179]

The Empire Divides (395 AD)

The Roman Empire was rich when Augustus became Emperor in 31 BC.[180] The state's financial resources were plentiful, both from the loot and plunder acquired during conquests as well as from gold and silver mines. Furthermore, the empire possessed rich agricultural lands, the

---

[172] Korn et al. (2006), Separation and preconcentration procedures for the determination of lead using spectrometric techniques: A Review, Talanta, 69, pp. 16–24.

[173] Ibid.

[174] Ibid.

[175] Byers, R., and Lord, E. (1943), Late Effects of Lead Poisoning on Mental Development, Am J Dis Child, 66(5), pp. 471–494.

[176] Korn et al. (2006), Separation and preconcentration procedures for the determination of lead using spectrometric techniques: A Review, Talanta, 69, pp. 16–24.

[177] Temin, P. (2006), The Economy of the Early Roman Empire, Journal of Economic Perspectives, Vol. 20, No. 1, pp. 133–151.

[178] Wilson, A. (2002), Machines, Power, and the Ancient Economy, Journal of Roman Studies. 92, pp. 1–32.

[179] Robinson, D. (1926), Roman Sculptures from Colonia Caesarea (Pisidian Antioch), The Art Bulletin, Vol. 9, No. 1, pp. 4–69.

[180] Morgan, J. (2012), The Roman Empire, Fall of the West: Survival of the East, AuthorHouse, Bloomington, USA.

army was well-trained and could be readily deployed, taxes were being collected, and work could be completed by slaves.[181] However, by the third century AD, agricultural production in the Roman Empire was beginning to decline. In the empire's North African domains, the Berbers were reclaiming their former domain and destroying crops.[182] A diversity of crops and foodstuffs needed by the empire were produced in Hispania. However, due to a decline in population because of the plague as well as increasing transport costs, the crops and foodstuffs could not be moved to parts of the Empire, where it was needed.[183] The Rhine-Danube frontier separated the lands controlled by the Roman Empire, to its south, and the lands of the Germanic tribes. Over the centuries, it became increasingly difficult for the Roman legions to keep back the Germanic hordes from the region's rich agricultural lands.[184] As a result, the Germanic hordes could destroy crops and occupy more and more land on an ever-increasing basis, and agricultural production began to decline in this region too. In conjunction with the increasing attacks by the Germanic hordes, the Roman Empire was itself going through a series of self-destructive civil wars. The result was that the legions were being drained of resources that could have been used to subdue the invaders on a sustained basis. The first of the Germanic peoples to enter the lands of the Roman Empire were the Goths, who crossed the Danube in 376 AD, fleeing the Huns who were in turn entering their lands from the Eurasian Steepes.[185] In 378 AD, the Goths inflicted a substantial defeat on the Empire's eastern army at the Battle of Hadrianopolis.[186] However, it was the western empire which was to suffer the most at the hands of the Goths who entered northern Italy in 401 AD.[187] The western Roman Empire's armies were stretched even further in 406 AD, when three more

---

[181] Ibid.

[182] Ibid.

[183] Ibid.

[184] Ibid.

[185] Ward-Perkins, B. (2005), The Fall of Rome and the End of Civilisation, Oxford University Press, Oxford.

[186] Ibid.

[187] Ibid.

Germanic tribes, the Vandals, the Sueves, and the Alans crossed the River Rhine into Roman-occupied Gaul.[188] From then on, the Romans were challenged in their own territories, with the Vandals even capturing Carthage, the Roman capital of north Africa, in 429 AD and 439 AD.[189]

The Roman Empire was also engaging in trade in silk, spices, and jewellery with the civilisations of the east, such as those on the Indian subcontinent. This had the effect of draining even more wealth from the Romans.[190] The Romans also made the situation worse when, in 395 AD, the empire was divided between east and west. The former, now known as the eastern empire acquired the protected and highly productive rich agricultural lands, while the western empire acquired the less productive and the less protected agricultural lands to the west.[191] Following the division of the Empire between east and west in 395 AD, the western empire collapsed 81 years later, in 476 AD.[192] It was in this year that the last Roman Emperor of the western empire, Romulus Augustulus, was ousted by a mercenary of Germanic origin, Odoacer. The title of Emperor of the western empire passed to Nespos. However, there is some ambiguity as to whether it was King Odoacer of Italy or Emperor Nespos of the western empire who ruled Italy.[193] Nevertheless, while Nespos was murdered in 480 AD, Odoacer held the power in Italy.[194] But the removal of Romulus Augustulus was the event that marked the end of the western empire. The military, economic, and political disintegration of the western empire resulted in the start of the Dark Ages of intellectual and material poverty in Europe.[195] It was a time during which no nation or civilisation was sophisticated enough to fill the shoes left by a vanishing

[188] Ibid.

[189] Ibid.

[190] Morgan, J. (2012), The Roman Empire, Fall of the West: Survival of the East, AuthorHouse, Bloomington, USA.

[191] Ibid.

[192] Ibid.

[193] Thompson, E. (1982), Romans and Barbarians – The Decline of the Western Empire, The University of Madison Press, Madison, USA.

[194] Ibid.

[195] Ward-Perkins, B. (2005), The Fall of Rome and the End of Civilisation, Oxford University Press, Oxford.

civilisation. As the end of the western empire, ancient Rome marked the start of the Dark Ages in Europe and also heralded the beginning of the Middle Ages and Feudalism in Europe.[196] The Germanic kingdoms which emerged in the west, in the aftermath of the end of the western empire, were better able to defend their borders at less of a cost than the cost to the western empire of defending its borders.[197] The lands of the north African coast had been the granary of the Roman Empire.[198] But, when the Vandals had taken north Africa from the Romans, they were managing agriculture efficiently—leading to greater prosperity than had been possible under the Romans.[199] However, when the Eastern Roman Emperor Justinian reconquered the region from the Vandals, the region's prosperity began to decline.[200] These two cases may suggest that the marginal return on social investment will be the larger, the lower the level of complexity of governance.[201] On the other hand, in contrast to the western empire, the eastern empire was far wealthier and, as a result, it was able to better finance the management of its complexity, in terms of the army, food production, administration, and construction. As a result, the eastern empire was able to survive longer simply because its marginal return on social investment was higher than that of the western empire.[202]

The eastern empire survived for nearly another 1000 years more than the western empire did. There were two factors which were most responsible for the collapse of the western empire and the continued survival of the eastern empire. The first was the decline in agricultural productivity. And the second was the difference in the possession of the most productive and well protected lands between the western empire and the eastern empire. In this case, both factors favoured the eastern empire rather than

[196] Ibid.

[197] Tainter, J. (1988), The Collapse of Complex Societies, Cambridge University Press, Cambridge.

[198] Davies, G. (2002), From Primitive and Ancient Money to the Invention of Coinage, 3000–600 BC, History of Money, University of Wales Press.

[199] Tainter, J. (1988), The Collapse of Complex Societies, Cambridge University Press, Cambridge.

[200] Ibid.

[201] Ibid.

[202] Ibid.

the western empire.[203] A decline in agricultural productivity resulted in less food to feed the people and the legions, as well as a decline in taxation.[204] This would mean that the state would have been unable to buy the foodstuffs that it could not produce, nor be able to pay its soldiers or hire mercenaries to protect its borders. Nevertheless, in the eastern empire, the Emperor Justinian did resort to heavy taxation to carry out an extensive building programme or reconquer the former lands of the western empire.[205] The eastern empire was far wealthier too, in comparison to the former western empire, and taxes could draw substantial financial resources from skimming the surpluses of the wealthy.[206]

References

Alcock, S. (2007), The Eastern Mediterranean, IN: The Cambridge Economic History of the Greco-Roman World, Scheidel, W., Morris, I., Saller, R. (Eds), Cambridge University Press, Cambridge.

Arena, V. (2012), Libertas and the Practice of Politics in the Late Roman Republic, Cambridge University Press, Cambridge.

Biagini, E. (1994), Roman Law and Political Control – from a Primitive Society to the Dawn of the Modern World, GeoJournal, 33.4, pp. 331–340.

Buttle, N. (2001), Republican Constitutionalism: A Roman Ideal, The Journal of Political Philosophy, Vol. 9, No. 3, pp. 331–349.

Byers, R., and Lord, E. (1943), Late Effects of Lead Poisoning on Mental Development, Am J Dis Child, 66(5), pp. 471–494.

Cherry, D. (2007), The Frontier Zones, IN: The Cambridge Economic History of the Greco-Roman World, Scheidel, W., Morris, I., Saller, R. (Eds), Cambridge University Press, Cambridge.

Cline, E. (2013), The Trojan War – A Very Short Introduction, Oxford University Press, Oxford.

[203] Morgan, J. (2012), The Roman Empire, Fall of the West: Survival of the East, AuthorHouse, Bloomington, USA.

[204] Ibid.

[205] Ibid.

[206] Derks, H. (2006), The Ancient Economy and its Countryside, The European Legacy, 11:5, pp. 541–546.

Conant, E. (1928), The Laws of the Twelve Tables, 13, St Louis L.Rev.231
Cornell, T. (1995), The Beginnings of Rome: Italy and Rome from the Bronze Age to the Punic Wars (c. 1000–264BC), Routledge, New York.
David, P. (1998), Common Agency Contracting and the Emergence of 'Open Science' Institutions. American Economic Review. 88:2, pp. 15–21.
Davies, G. (2002a), From Primitive and Ancient Money to the Invention of Coinage, 3000–600BC, History of Money, University of Wales Press.
Davies, G. (2002b), The Development of Greek and Roman Money, 600BC–610AD, History of Money, University of Wales Press.
Derks, H. (2006), The Ancient Economy and its Countryside, The European Legacy, 11:5, pp. 541–546.
Dmitriev, S. (2004), Traditions and innovations in the Reign of Aurelian, Classical Quarterly, 54.2, pp. 568–578.
Duncan, M. (2016), The History of Rome – Volume 1: The Republic, Herodotus Press.
Fenn, P. (1925), Justinian and the Freedom of the Sea, The American Journal of International Law, Vol. 19, No. 4, pp. 716–727.
Flower, H. (2014), Preface, IN The Cambridge Companion to The Roman Republic, Flower, H. (Eds), Cambridge University Press, New York.
Giardina, A. (2007), The Transition to Late Antiquity, IN: The Cambridge Economic History of the Greco-Roman World, Scheidel, W., Morris, I., Saller, R. (Eds), Cambridge University Press, Cambridge.
Goodman, M. (1997), The Roman World, *44BC–AD 180.* London, Routledge.
Gruen, E. (1974), The Last generation of the Roman Republic, University of California Press, Berkeley.
Hopkins, K. (1978), Conquerors and Slaves, Cambridge University Press, Cambridge.
Keaveney, S. (2005), The Last Republican: Sulla, Routledge, New York.
Keppie, L. (1998), The Making of the Roman Army: From Republic to Empire, University of Oklahoma Press, Norman.
Kessler, D., and Temin, P. (2005), Money and Prices in the Early Roman Empire, Working Paper 05–11, MIT.
Kessler, D., and Temin, P. (2007), The organisation of the Grain Trade in the early Roman Empire, Economic History Review, 60, 2, pp. 313–332.
Korn et al. (2006), Separation and preconcentration procedures for the determination of lead using spectrometric techniques: A Review, Talanta, 69, pp. 16–24.

Leveau, P. (2007), The Western Provinces, IN: The Cambridge Economic History of the Greco-Roman World, Scheidel, W., Morris, I., Saller, R. (Eds), Cambridge University Press, Cambridge.

Lintott, A. (1999), The Constitution of the Roman Republic, Oxford University Press, Oxford.

Lomas, K. (2017), The Rise of Rome, 1000 BC to 264 BC, Profile Books Ltd, London.

Mackay, C. (2004), Ancient Rome: A Military and Political History, Cambridge University Press, Cambridge.

McCarty, N. (2008), Rome – The Greatest Empire of the Ancient World, The Rosen Publishing Group, Inc, New York.

Millar, F. (1986), Politics, Persuasion and the People before the Social War, Journal of Roman Studies, lxxvi, pp. 1–11.

Moatti, C. (2015), The Birth of Critical Thinking in Republican Rome, Cambridge University Press, Cambridge.

Mouritsen, H. (2004), Plebs and Politics in the Late Roman Republic, Cambridge University Press, Cambridge.

Nicolet, C. (1980), The World of the Citizen in Republican Rome, University of California Press, Berkeley.

Frank, P. (2004), Philosophy of Science: The Link Between Science and Philosophy, Dover Publications Inc, New York.

Morgan, J. (2012), The Roman Empire, Fall of the West: Survival of the East, AuthorHouse, Bloomington, USA.

Raaflaub, K. (2006), Between Myth and History: Rome's Rise from Village to Empire (the eighth century to 264), IN: A Companion to the Roman Republic, Rosenstein, N., and Morstein-Marx, R. (Eds), Blackwell Publishing, Oxford, USA.

Robinson, D. (1926), Roman Sculptures from Colonia Caesarea (Pisidian Antioch), The Art Bulletin, Vol. 9, No. 1, pp. 4–69.

Rosenstein, N. (2009), War, State Formation, and the Evolution of Military Institutions in Ancient China and Rome, IN Rome and China: Comparative Perspectives on Ancient World Empires, Scheidel, W. (Ed), Oxford University Press, Oxford.

Sahay, A. (2005), Review Essay: Managing Innovation and Technology, The Journal of Business Perspectives, Vol. 9, No. 4.

Scott-Kilvert, I. (1979), The Rise of the Roman Empire, Polybius; translated by Ian Scott-Kilvert; selected with an Introduction by F.W. Walbank, Penguin Classics, Harmondsworth.

Shapiro, M. (1981), Courts: A Comparative and Political Analysis, The University of Chicago Press, Chicago.

Tainter, J. (1988), The Collapse of Complex Societies, Cambridge University Press, Cambridge.

Temin, P. (2001), A Market Economy in the Early Roman Empire, The Journal of Roman Studies, Vol. 91, pp. 169–181.

Temin, P. (2004a), Financial Intermediation in the Early Roman Empire, The Journal of Economic History, Vol. 64, No. 3, pp. 705–733.

Temin, P. (2004b), The Labour Market of the Early Roman Empire, Journal of Interdisciplinary History, XXXIV:4, pp. 513–538.

Temin, P. (2006), The Economy of the Early Roman Empire, Journal of Economic Perspectives, Vol. 20, No. 1, pp. 133–151.

Thapar, R. (1988), Epigraphic evidence and some Indo-Hellenistic contacts during the Mauryan period, in Maity, S.K., and Thakur, U. (eds.), *Indological Studies, Prof. D. C. Sircar Commemoration Volume,* New Delhi, pp. 15–19.

Thapar, R. (1992), Black Gold: South Asia and the Roman Maritime Trade, South Asia: Journal of South Asian Studies,15:2, pp. 1–27.

Thompson, E. (1982), Romans and Barbarians – The Decline of the Western Empire, The University of Madison Press, Madison, USA.

Vervaet, F. (2004), The 'Lex Valeria' and Sulla's Empowerment as Dictator (82–79BCE), Cahiers du Centre Gustave Glotz, Vol. 15, pp. 37–84.

Ward-Perkins, B. (2005), The Fall of Rome and the End of Civilisation, Oxford University Press, Oxford.

Wilson, A. (2002), Machines, Power, and the Ancient Economy, Journal of Roman Studies. 92, pp. 1–32.

Wiseman, T. (1998), Roman Republic Year One, Greece & Rome, Vol. XLV, No. 1.

7

The Chola Dynasty: 350 BC to 1279 AD

Introduction

The earliest written indication of the states of southern India was by the King of Kalinga in the second century BC, in the form of inscriptions celebrating a victory in a battle against a Tamil confederacy in 155 BC.[1] Kautilya, advisor to Chandragupta, the first Mauryan Emperor noted—in the classical work, the Arthashastra—in the fourth century BC that the southern land route was easily travelled.[2] These facts would suggest that the first Tamil kingdoms were established around the fourth century BC. The roots of the Chola dynasty lie in 850 AD, when the Chola chieftain Vijayalaya captured the city of Tanjavur with the approval of the Pallava dynasty from the Muttaraiyars.[3] Furthermore, the Velirs of Kodumbalur provided military assistance to Vijayalaya.[4] Nevertheless,

[1] Kainikara, S. (2016), From Indus to Independence: A Trek Through Indian History, Volume II, The Classical Age, Vij Books India Pvt Ltd., New Delhi.

[2] Danino, M. (1988), Vedic Roots of Early Tamil Culture, Vivekananda Kendra Patrika, Vol. 40, No. 1, 79th Issue.

[3] Schmidt, K. (1995), An Atlas and Survey of South Asian History, Routledge, New York.

[4] Ibid.

S. Ramesh, *The Rise of Empires*, https://doi.org/10.1007/978-3-030-01608-1_7

the descendants of the Chola royal line, from its preeminent time, served as subservient princes to the Pallavas and sought to reassert their power at a time when Chola power was on the wane.[5] In this case, the latter suggests that, when the Pallavas were away fighting the Pandyans in the north, the Chola Prince Aditya challenged the commander of a Pallava army and won. Aditya was succeeded by King Parantika 1 (907 AD–946 AD). The latter consolidated the power of the Chola dynasty in southern India for the next few centuries.[6] Parantika 1 defeated the Pandyans and absorbed their lands into that of the Chola dynasty, as well as militarily destroying the power of the Pallavas. Nevertheless, despite the successes of Parantika 1, he also suffered a serious military defeat at the hands of the Rashtrakuta King, Krishna 3 and, compared to Parantika 1, his next five successors were militarily weaker.[7] Therefore, the Cholas were a minor dynasty in southern India until they supplanted the other dynasties of the region, their rule not only encompassing southern India, but also other parts of Asia.[8] This geographical expansion had been achieved not only due to military power, but also through the development of a ruling ideology which took account of the pre-existing chiefly customs, and at the same time increased the hold of the Chola kings over the Tamil lands.[9] The latter suggests that the Chola Kings maintained power through a process of exchange and accessibility. The exchange of women between the King and his local chieftains served to achieve 'family' loyalty. Moreover, while the King was being divine because he maintained a Siva cult at his place of worship, protected the brahmans, and was the epitome of dharma, his spirituality was not a monopoly.[10] It would become

[5] Kulke, H., and Rothermund, D. (1998), A History of India, Routledge, New York.

[6] Chaurasia, R. (2008), History of Ancient India: Earliest Times to 1200 AD, Atlantic Publishers & Distributors (p) Ltd., New Delhi.

[7] Ibid.

[8] Heitzman, J. (1991), Ritual Polity and Economy: The Transactional Network of an Imperial temple in Medieval South India, Journal of the Economic and the Social History of the Orient, Vol. 34, No. ½, pp. 23–54.

[9] Stein, B. (1985), Politics, Peasants and the Deconstruction of Feudalism in Medieval India, in Feudalism and Non-European Societies, Byres, T., and Mukhia, H. (Eds), Frank Cass and Company Limited, London.

[10] Ibid.

accessible to local village chieftains who accepted the suzerainty of the Chola King. Thus, the process of exchange and accessibility could be a way in which the Chola kings were able to maintain their control over Southern India. Because of the rising power of the Cholas, Tamil merchants were able to engage in maritime trade with the Middle East, Southeast Asia, and China. These maritime links may also have extended to include Japan.[11] Military, economic, and political power allowed the Cholas to remain a dominant power for a few hundred years.[12] The Chola dynasty was also mentioned in the chronicles of China's Song dynasty, the Songshi, and was referred to as Zhunian.[13] The Songshi also cites a visit to the Song court in China by a Chola ambassador in 1077.[14] The southern states of India lie below the Deccan Plateau, and this region is often referred to as the Tamil country. The early history of the South Indian states in the period BC and in the early part of the period AD are celebrated in the Sangam, the earliest form of Tamil literature.[15] The Sangam was a gathering of scholars and poets under royal patronage. Supposedly, there were three Sangams, with the last Sangam taking place in the city of Madurai in the fifth century BC; and the Sangam period encompassed 350 BC to 300 AD. This approximates to the early Chola period, but there is no evidence of any link between the early Chola period and the new Chola dynasty, which reigned from 850 AD to 1279 AD.[16] This period can be broken down into four sections: rise and consolidation (850 AD to 985 AD), imperial expansion (986 AD to 1070

[11] DeRouen, K., and Heo, U. (2005), Defense and Security: A Compendium of National Armed Forces and Security Policies: Vol. 1, ABC-CLIO, Santa Barbara, CA.

[12] Sastri, K. (1958), A History of South India from Pre-historic Times to the fall of Vijayanagar, Oxford University Press, London.

[13] Mahalakshmi, R. (2016), Chola Empire, IN MacKenzie, J. (Ed), The Encyclopedia of Empire, John Wiley & Sons Ltd., London.

[14] Sen, T. (2010), The Military Campaigns of Rajendra Chola and the Chola-Sri Vijaya-China Triangle. In H. Kulke et al. (Eds.), Nagapattinam to Suvarnadwipa: Reflections on the Chola Naval Expeditions to Southeast Asia: 61–75.

[15] Kainikara, S. (2016), From Indus to Independence: A Trek Through Indian History, Volume II, The Classical Age, Vij Books India Pvt Ltd., New Delhi.

[16] Mahalakshmi, R. (2016), Chola Empire, IN MacKenzie, J. (Ed), The Encyclopedia of Empire, John Wiley & Sons Ltd., London.

AD), a struggle to maintain power (1071 AD to 1178 AD), and economic and political decline (1179 AD to 1279 AD).

The Sangam Period

It was during the Sangam period—200 BC to 200 AD—that the earliest classical Tamil writings emerged. Three Sangams—or gathering of scholars, poets, and writers—were held over a 6000-year period, with the first Sangam being held in around 6000 BC. Each Sangam had a period of about 2000 years. The Sangam classical Tamil writings vary in length, from a few lines to hundreds of lines, written by under five-hundred authors.[17] Moreover, the latter suggests that the Sangam can be decomposed into nine schematic anthologies, which includes the Narrinai, the Kurundogai, the Aingurunuru, the Padirruuppattu, the Paripadal, the Kalottogai, the Ahananuru, the Purananuru, and the Pattuppattu. The time in which these anthologies developed coincides with the emergence of the Tolkappiyam, the epic treatise on Tamil grammar. Despite the existence of some works of Tamil literature, others, such as the translation of the Mahabharata in the period encompassing early Pandyan rule, have been lost in the mists of time.[18] The importance of the Sangam anthologies is that, while they may be a mixture of fact and faction, they nevertheless provide a picture of daily social lives.[19] Moreover, the Sangam literature also provides evidence of trade between the Tamil states and Greece and Rome, which is corroborated by the works of the Greek writers Strabo and Ptolemy and the Roman writer Pliny.[20] The evidence of trade between the South Indian states and Greece and Rome in ancient times, depicted in the Sangam literature as well as in the literature of Rome and Greece, has been supported by archaeological evidence, such as the Greco-Roman port at Arikamedu, Pondicherry, in southern India,

[17] Kainikara, S. (2016), From Indus to Independence: A Trek Through Indian History, Volume II, The Classical Age, Vij Books India Pvt Ltd., New Delhi.

[18] Ibid.

[19] Ibid.

[20] Ibid.

and the discovery of Roman coins from the first and second century AD.[21] Greco-Roman trade with southern India increased after the Greek navigator Hippalus discovered how to make the best use of the monsoon winds in order to traverse the ocean from west to east.[22] According to the latter, better knowledge of the geography of the South Indian coast also helped the Greco-Romans to sail to southern India. However, other South Indian ports such as Alagankulam and Puhar also facilitated Greco-Roman trade, with Roman colonies existing at each port including the inland city of Madurai.[23] The Tamils called the Greco-Romans 'Yavanas', and the 'Yavanas' were in demand by the Tamils to work for them as craftsmen, carpenters, and traders.[24] The Romans consumed more than they produced, and much of Rome's gold bullion went to South India.[25] Nevertheless, the Arabs were the first to trade with South Indian merchants, and were only superseded by the Greek subjects of the Roman Empire in the early Christian period.[26] The trade in spices, precious stones, and muslins thrived from 31 BC, to the time of Nero in 68 AD. However, this trade declined after his death, although it continued until the third century. The reason for the decline in trade between Rome and South India was that the Emperor Vespasian, who succeeded Nero, curbed the appetite of the wealthy Romans by restricting the import of goods from South India. This reduced the outflow of Roman gold.[27] This outflow had amounted to 550 million sesterces a year.[28] The ancient link

[21] Kainikara, S. (2016), From Indus to Independence: A Trek Through Indian History, Volume II, The Classical Age, Vij Books India Pvt Ltd., New Delhi.

[22] Howard, M. (2012), Transnationalism in Ancient and Medieval Societies: The Role of Cross-Border Trade and Travel, McFarland & Company, North Carolina.

[23] Mukund, K. (2012), The World of the Tamil Merchant: Pioneers of International Trade, Penguin Books, Haryana, India.

[24] Ball, W. (2016), Rome in the past: The Transformation of an Empire, 2nd Edition, Routledge, Oxon.

[25] Mukund, K. (2012), The World of the Tamil Merchant: Pioneers of International Trade, Penguin Books, Haryana, India.

[26] Mendis, G. (2005), The Early History of Ceylon: And Its Relations With India and Other Foreign Countries, Asian Educational Services, New Delhi, Chennai.

[27] Mukund, K. (2012), The World of the Tamil Merchant: Pioneers of International Trade, Penguin Books, Haryana, India.

[28] Thapar, R. (1990), A History of India: Volume 1, Penguin Books, London.

between the Greeks and the Tamils was so strong that some spices such as pepper and ginger are in fact derived from Tamil words.[29]

The different regions of southern India were ruled by three dynasties in ancient times. **Wilkinson (1996)** refers to this regional power configuration as one of 'hegemonic unipolarity' in which there is either one superpower, no great powers, or three or more local powers. While the three dynasties—the Cheras, the Cholas, and the Pandyas—were the dominant powers, smaller regions were ruled by local chieftains or Vallals who, while they owed allegiance to a dynasty, were able to exercise unilateral authority over the peoples in their jurisdiction. The first recorded king in the ancient Chola dynasty is Ilanjetcenni, who is descended from an earlier unknown Chola king who is reported to have had the wind do his bidding. This may be a reference to the early Chola maritime excursions.[30] The King who followed Ilanjetcenni was Karikala Peruvalathaan, his son, who ruled around 270 BC. Karikala won many notable battles, including the Battle of Veni, in which he defeated 11 minor kings or chieftains, and the Battle of Vahaiparandalai, in which he defeated 9 minor kings. In another battle, Karikala defeated the combined armies of the Pandya and the Chera kings.[31] Karikala maintained his capital at Kaveri Pattinam, and from here he instigated policies that prevented migration from his lands, the resettlement of forested lands, the development of flood banks along the River Kaveri, and the development of irrigation canals, water tanks, and dams to sustain agriculture.[32] One of these dams was the Kallanai Dam across the Kaveri River, which is still in use today. The Chola king who took it to the height of its powers was Rajaraja 1 (985 AD–1014 AD) at the beginning of the eleventh century. Rajaraja 1 overcame the military strength of dynasties such as the Pandyan kingdom, the island of Ceylon, the Western Chalukyas, as well as the Eastern Chalukyas. In order to celebrate his military victories, Rajaraja 1 built a major temple

[29] Mendis, G. (2005), The Early History of Ceylon: And Its Relations With India and Other Foreign Countries, Asian Educational Services, New Delhi, Chennai.

[30] Kainikara, S. (2016), From Indus to Independence: A Trek Through Indian History, Volume II, The Classical Age, Vij Books India Pvt Ltd., New Delhi.

[31] Ibid.

[32] Ibid.

complex, the Rajarajesvara Temple, in his capital city of Tanjavur.[33] At the time, the Rajarajesvara Temple was the largest in the world and, having taken seven years to build, it also exhibited a number of architectural innovations.[34]

The Chola State

In pre and early Chola periods, peasant localities were organised into units, each of which were referred to as a 'nadu'. Together with brahman elements playing a central role in daily life, the nadu represented a political system.[35] Nevertheless, the nadu also represented a specific land area in which agricultural production was organised.[36] The nadus themselves were organised into 'Periyanadu' in the thirteenth century and 'Mandalam' in the eighteenth century. The 'Periyanadu' was in fact a guild of agriculturalists, producers, and traders in agricultural commodities.[37] It has also been noted that, in the 'Periyanadu', all four castes were represented, and the 'Periyanadu' acted in conjunction with other supra-local institutions in order to organise production.[38] Moreover, the same association between these institutions allowed for the support of trading and merchants in the years following the collapse of the Chola Empire in the 13th dynasty.[39] The brahman elements of the nadus incorporated learned individuals

[33] Heitman, J. (1997), Gifts of Power: Lordship in an Early Indian State, Oxford University Press, Delhi.

[34] Reddy, V. (2009), Temples of South India, Guyan Publishing House, New Delhi.

[35] Stein, B. (1977), Circulation and the Historical Geography of Tamil Country, The Journal of Asian Studies, Vol. 37, No. 1, pp. 7–26.

[36] Karashima, N., Subbarayalu, Y., and Shanmugam, P. (2008), Nagaram During the Chola and Pandyan Period: Commerce and Towns in Tamil Country AD 850–1350, The Indian Historical Review, Volume XXXV, No. 1, pp. 1–33.

[37] Champakalakshmi, R. (1996), Trade, Ideology and Urbanization in South India 300 BC to AD 1300, Oxford University Press, Delhi.

[38] Krishnan, K.G. (1982), Chittirameml Mip-periyanamdDu – An Agricultural Guild of Medieval Tamil Nadu. Journal of the Madras University, Vol. 54:1.

[39] Karashima, N., and Subbarayalu, Y. (2004), The Emergence of the Periyanadu Assembly in South India during the Chola and Pandyan Periods, International Journal of Asian Studies, 1,1, pp. 87–103.

with a knowledge of Sanskrit, centred around temples and the worship of puranic gods, societal norms associated with purity and respectability derived from the Sanskrit shastras which emphasised the support of legitimate kingship, and stressed the righteousness of vegetarianism.[40] The building of the Rajarajesvara Temple was both an act of worship as well as a symbol of the institutional power of the King.[41] The temple was dedicated to the Hindu god Shiva, with the King and other royal notables making substantial donations for its construction, maintenance, and daily functioning. Due to the importance of the Rajarajesvara Temple, priests, temple dancers, and other types of workers necessary for the functioning of the temple were drawn from smaller temples throughout Southern India. At the time of its construction, the Rajarajesvara Temple was the tallest site of worship in South Asia, rising to 190 ft.[42] The height of the tower was a physical manifestation of Rajaraja's skills, abilities, and power.[43] The message of the temple to citizens was clear. The creator of the temple was the leading citizen, the King, who had the role of maintaining stability in the cosmos. In other words, to ensure stability and peace in the country so that the citizens could go by their daily lives. Making donations to the temple evidenced piety and humility to the Lord Shiva and, in a way, to the King, as the protector of the cosmos. The façade of the temple exemplifies traditional architectural practice and innovation.[44] This was because the traditional tiers of the façade were kept, but the proportions were such as to make the physical appearance wider but smaller.[45] Rajaraja's successor built the Gangaikondachokpuram

[40] Stein, B. (1977), Circulation and the Historical Geography of Tamil Country, The Journal of Asian Studies, Vol. 37, No. 1, pp. 7–26.

[41] Heitzman, J. (1991), Ritual Polity and Economy: The Transactional Network of an Imperial temple in Medieval South India, Journal of the Economic and the Social History of the Orient, Vol. 34, No. ½, pp. 23–54.

[42] Ibid.

[43] Talwai, P. (2010), Praying through Politics, Ruling Through Religion: The Rajarajeswaram as an instrument of Economic and Political Unification in the Chola Empire, UC Berkeley, scholarship, https://escholarship.org/uc/item/70v1v9dh

[44] Prichard, P. (1995), Tanjavur Brhadisvara: An Architectural Study, Indira Gandhi Centre for the Arts, New Delhi.

[45] Ibid.

temple. Other innovations also occurred in metallurgy, and the production of South Indian medieval icons such as the dance of the Lord Shiva.[46] According to the latter, the medieval icons were produced using the first used high-tin beta bronzes. Furthermore, high-tin delta bronze mirrors were also produced.[47] The number of inscriptions to Temples in general, also increased during the Chola period from the ninth century AD onwards.[48] Chola era sculpture also showed innovation in showing more humanism.[49,50,51]

While, the temple architecture was both a religious and political symbol, the temple itself also held economic significance.[52] This was because of the high level of employment it created in the context of teachers, dancers, watchmen, shepherds, musicians, tailors, carpenters, lamplighters, and actors. The military also had a role in setting up icons, maintaining lamps and making donations of ornaments.[53] The construction and the daily functioning of the temple can be seen as a way in which Rajaraja 1 sought to legitimise and sustain his power in the face of the people.[54] In other words, Rajaraja I pivoted the political and economic power of his empire at the Rajarajeswaram Temple.[55] In general, temples in southern

[46] Srinivasan, S. (1998), The use of tin and bronze in prehistoric southern Indian metallurgy, *Journal of Metals*, 50(7), 44–9.

[47] Ibid.

[48] Karashima, N. (1996), South Indian Temple Inscriptions: A New Approach to their study, South Asia: Journal of Asian Studies, 19, 1, 1–12.

[49] Irwin, J. (1948), South Indian Figure Sculpture, 7th–10th Centuries AD, Journal of the Royal Asiatic Society.

[50] Lippe, A. (1960), The Sculpture of Greater India, The Metropolitan Museum of Art Bulletin, New Series, Vol. 18, No. 6, pp. 177–192.

[51] Barrett, D. (1964), An Early Cola Lingodbhavamurti, The British Museum Quarterly, Vol. 28, No. 1/2, pp. 32–39

[52] Venkataraman, B. (1985), Rajarajesvaram: The Pinnacle of Chola Art, Mudgala Trust, Madras.

[53] Ibid.

[54] Heitzman, J. (1991), Ritual Polity and Economy: The Transactional Network of an Imperial temple in Medieval South India, Journal of the Economic and the Social History of the Orient, Vol. 34, No. ½, pp. 23–54.

[55] Talwai, P. (2010), Praying through Politics, Ruling Through Religion: The Rajarajeswaram as an instrument of Economic and Political Unification in the Chola Empire, UC Berkeley, scholarship, https://escholarship.org/uc/item/70v1v9dh

India were the centres of economic, social, and political life.[56] The size and splendour of the Rajarajeswaram Temple gave it a greater level of significance in this context. The temple was not only funded by the interest from deposits, but also by the collection of taxes. The main source of income to pay the wages of temple workers was through a system of land grants, the devadana. Property rights during the time of the Cholas were not in the same context as that of a capitalist economy.[57] All land was owned by the King, and there was only a 'shift' of land possession through sales.[58] This perhaps excluded temples. Land sales were of four types.[59] These included the sale of cultivator's land, the sale of the land of condemned officials, the sale of village land, and the sale of land by temples. There were two types of devadana. The first type allowed the temple to retain rights to revenues from the land, while the cultivator retained possession of the land. The second type of devadana conferred ownership of the land to the temple.[60] However, according to the latter, the tenant cultivators were responsible for producing the crops. The tax due from the cultivated land was collected by the merchants, Rajaraja again using a pre-existing social condition, with the collected revenues going to the temple.[61] Nevertheless, the produce of the land would be shared between the temple, the state, and by the cultivator.[62] However, this traditional system of dividing the produce of the land was changed by Rajaraja so

[56] Mukund, K. (2012), The World of the Tamil Merchant: Pioneers of International Trade, Penguin Books, Haryana, India.

[57] Subrahmanyam, S. (1990), The Political Economy of Commerce: Southern India, 1500–1650, Cambridge University Press, Cambridge.

[58] Gough, K. (1980), Modes of Production in Southern India, Economic and Political Weekly, Vol. 15, No. 5/7.

[59] Sastri, K. (1958), A History of South India from Pre-historic Times to the fall of Vijayanagar, Oxford University Press, London.

[60] Venkatachari, K.K.A. (1987), Rajaraja the Great: Seminar Proceedings, Ananthacharya Indological Research Institute, Bombay.

[61] Talwai, P. (2010), Praying through Politics, Ruling Through Religion: The Rajarajeswaram as an instrument of Economic and Political Unification in the Chola Empire, UC Berkeley, scholarship, https://escholarship.org/uc/item/70v1v9dh

[62] Karashima, N., Subbarayalu, Y., and Shanmugam, P. (2008), Nagaram During the Chola and Pandyan Period: Commerce and Towns in Tamil Country AD 850–1350, The Indian Historical Review, Volume XXXV, No. 1, pp. 1–33.

that the temple and the state would receive more revenues. It should be noted that the 'share' system was very similar to the modern-day concept of the 'mutual' in the context of increasing incentives to work harder, and thus to produce more.[63]

The devadana was a system by which Rajaraja could maintain political control of not only his kingdom, but also over his conquered lands.[64] The importance of specific geographical regions to tax collection increased, the greater the distance of the region from the imperial capital Palaiyarv.[65] For it was in Palaiyarv that royal rituals took place, donations received, and royal residences persisted. On the other hand, Tanjavur can be seen as the informal capital, whose construction and development would facilitate economic activity and economic development.[66] This economic activity and economic development was facilitated by the use of coinage. Coinage has been used in India to facilitate transactions for over 2000 years; and the use of coinage in India has preceded the use of coinage in Britain by a few centuries.[67] According to the latter, because the Southern Indian dynasties were able to prevail over the Islamic invasion of the north, traditional Indian coin design continued with the distinctive fish design on coins in Chola territory. Other excavations have produced early period (circa fourth/fifth century) Chola copper coins with depictions of the elephant, the ritual umbrella, and Hindu symbols such as the Srivatsa and the Swastika.[68] Furthermore, despite the Muslim incursions into northern India, southern India under the Cholas and the other dynasties were able to remain Hindu in character.[69]

---

[63] Mukund, K. (2012), The World of the Tamil Merchant: Pioneers of International Trade, Penguin Books, Haryana, India.

[64] Vasudevan, G. (2003), The Royal Temple of Rajaraja: An Instrument of Imperial Chola Power. New Delhi: Abhinav.

[65] Heitzman, J. (1987), Temple Urbanism in Medieval South India, Journal of Asian Studies, Vol. 46, No. 4, pp. 791–826.

[66] Ibid.

[67] Cribb, J. (2003), The Origins of the Indian Coinage Tradition, South Asian Studies, 19, 1, pp. 1–19.

[68] Danino, M. (1988), Vedic Roots of Early Tamil Culture, Vivekananda Kendra Patrika, Vol. 40, No. 1, 79th Issue.

[69] Basham, A. (1954), The Wonder that was India, Sidgwick and Jackson, London.

The Brahmadeya, Ur, and the Nagaram were the towns and the villages that were included in the nadu. However, they may also be thought of in the context of democratic assemblies that presided over the governance of villages and towns.[70] According to the latter, these institutions predated the ascendancy of the Cholas and the other South Indian dynasties. The brahmadeya were the villages granted to the brahmans. As these brahman settlements received money deposits, they were the centres of development and of capital accumulation.[71] The Ur were the villages controlled by landowners and by the temples. The Nagaram represented a town, with a very similar word in Sanskrit, nagara. The Nagaram can be thought of as a corporate institution due to its exclusive membership of merchants.[72] Moreover, the Nagaram represented sites of economic density, because they were the abode of the merchants who were vital for commerce, trade, and urbanisation.[73] The merchants of the Nagaram were called 'Nagarattar' and were themselves organised into local guilds. The merchant guilds which traded abroad were able to buy the resources, such as ships, horses, donkeys, carts, and even soldiers to protect their goods, they needed to do so.[74] Furthermore, according to the latter, the merchant guilds were responsible for governing parts of towns on behalf of the King. And they also paid for the building of temples that they also managed on a day-to-day basis. The number of Nagarams varied from one nadu to another. And while the Chola kings encouraged the establishment of Nagarams in each nadu, there was no official economic policy to do so. However, the rising prominence of the Nagaram may contradict this view. Therefore, it may have been the case that one Nagaram may

[70] Mukund, K. (2012), The World of the Tamil Merchant: Pioneers of International Trade, Penguin Books, Haryana, India.

[71] Heitzman, J. (1987), Temple Urbanism in Medieval South India, Journal of Asian Studies, Vol. 46, No. 4, pp. 791–826.

[72] Mukund, K. (2012), The World of the Tamil Merchant: Pioneers of International Trade, Penguin Books, Haryana, India.

[73] Hall, K. (1980), Trade and Statecraft in the Age of the Cholas, Abhinav Publications, New Delhi.

[74] Gough, K. (1980), Modes of Production in Southern India, Economic and Political Weekly, Vol. 15, No. 5/7.

have served several nadus.[75] Nevertheless, unlike the Nagaram, the Brahmadeya were important for local government, and so, there was an official policy to develop at least one Brahmadeya per nadu.[76] However, the Nagarams were placed under the strict monitoring of the state.[77] This may have been due to their commercial importance and the tax revenues arising from trade. Thus, a town (Nagaram) in the Chola state did not enjoy the same level of freedom as did the medieval towns in Europe and Japan.[78] And the merchants or nagarattars in the Nagarams co-operated to a greater extent than did their counterparts in Europe. The nagarattars were also landowners, and any donations they made to temples were tax free.[79] The interest from the donations would be used to pay for services such as the materials needed to bathe the temple idols or repair temple structures. Typically, the interest would accrue as a percentage of specific commodities sold by merchants based on the money deposited with them.[80] The latter suggests that Nagarattars owned land, and they paid taxes on the land they owned to the state but controlled agricultural production on the same land. The most important Nagarattars or merchants in the Nagarams were the Viyapari (general merchants), the Sankarapadi (oil merchants), and the Saliya (weavers). Trading urban centres (Nagaram)as well as the local fiefdoms of chieftains enjoyed a degree of self-government. Moreover, the Chola kings franchised merchants by granting special charters which exempted them from either punitive or

[75] Karashima, N., Subbarayalu, Y., and Shanmugam, P. (2008), Nagaram During the Chola and Pandyan Period: Commerce and Towns in Tamil Country AD 850–1350, The Indian Historical Review, Volume XXXV, No. 1, pp. 1–33.

[76] Karashima, N. (1984), South Indian History and Society: Studies from Inscriptions, AD 850–1800, OUP, New Delhi.

[77] Karashima, N., Subbarayalu, Y., and Shanmugam, P. (2008), Nagaram During the Chola and Pandyan Period: Commerce and Towns in Tamil Country AD 850–1350, The Indian Historical Review, Volume XXXV, No. 1, pp. 1–33.

[78] Ibid.

[79] Ibid.

[80] Karashima, N., Subbarayalu, Y., and Shanmugam, P. (2008), Nagaram During the Chola and Pandyan Period: Commerce and Towns in Tamil Country AD 850–1350, The Indian Historical Review, Volume XXXV, No. 1, pp. 1–33.

custodial sanctions.[81] During the Chola period, trade and craft specialisation flourished due to the highly productive agricultural sector of the economy, which had been facilitated by the Chola state.[82] Nevertheless, in the Chola state, the King was not above the law, as interpreted by the brahmans, and the people were encouraged to overthrow the King if he did violate the law.[83] This was in contrast to China, where the Emperor was considered to be divine, and the centre of power. China then emerged as a stronger state than did India over the centuries. From a historical perspective, this led to China retaining its sovereignty, while only temporarily ceding parts of its territory to the European powers in the nineteenth century. However, India was completely absorbed by the British into the British Empire. Therefore, China's Emperor-oriented central rule acted as a force to unify the whole country. However, while this gave rise to the wars, such as the warring states period and the Spring and Autumn wars, this was not experienced in India's chronology because, before the advent of the British, India was never a unified state.[84] The roots of the institutional differences in power in India and China allowed both countries to evolve different forms of governance in the twentieth century. India inherited the western democratic model from its coloniser, although elements of democratic suffrage were exhibited in its ancient history. China, after a brutal foreign invasion, entered into a civil conflict which only ended with the victory of the Communist Party in 1949, resulting in one party rule, with a democratic element, ever since.

Increased trade facilitated craft specialisation to be concentrated in urban centres which also became a focal point for production. Local guilds of merchants specialising in the trade of a specific commodity emerged, and guilds of foreign merchants, mostly Arabs, but perhaps including Greeks and Romans, also existed. Karikala was also a follower

[81] Spencer, G. (1976), The Politics of Plunder: The Cholas in Eleventh-Century Ceylon, The Journal of Asian Studies, Vol. 35, No. 3, pp. 405–419.

[82] Mahalakshmi, R. (2016), Chola Empire, IN MacKenzie, J. (Ed), The Encyclopedia of Empire, John Wiley & Sons Ltd., London.

[83] Mukund, K. (2012), The World of the Tamil Merchant: Pioneers of International Trade, Penguin Books, Haryana, India.

[84] Ibid.

of the Vedic religion, which had permeated through the Deccan Plateau from the Aryan invaders in the north.[85] Under Chola rule, the formation of merchant guilds specialising in specific artisan sectors of the economy was also more favoured then before. The Chola Empire capitalised on the economic and maritime networks left by the Pallava empire following its demise in the eighth century AD, advancing them by using the power of the merchant guilds.[86] Increased trade between southern India, the Middle East, Southeast Asia, and China facilitated the rapid development of the weaving and dyeing industries, in southern India, with the introduction of the loom in the eleventh century and the spinning wheel in the thirteenth century.[87] This was 600 years earlier than the introduction of the spinning wheel at the start of the industrial revolution in England in the latter part of the eighteenth century. The thirteenth century also saw a rise in South Indian exports of printed and dyed textiles to the Middle East and Southeast Asia.[88,89] Furthermore, the merchant associations not only controlled the South Indian non-agrarian economy, but also internal and external trade from areas controlled by the Cholas.[90] Due to the monopoly power they held, the merchant guilds were able to accumulate sufficient wealth and patronise and support temples, as well as acting on behalf of the Chola state in order to collect taxes. However, while there was trade and agriculture, aspects of production were limited to textiles and dyes. The small composition of production in the Chola economy may have been due to the thinly distributed population, with pockets concentrated in the fertile deltas and valleys, in conjunction with

---

[85] Kainikara, S. (2016), From Indus to Independence: A Trek Through Indian History, Volume II, The Classical Age, Vij Books India Pvt Ltd., New Delhi.

[86] Christie, J. (1998), The medieval Tamil-language inscriptions in Southeast Asia and China, JSEAS, 29, 2, 257–8, Sastri, History, pp. 132–3.

[87] Abraham, M. (1988), Two Medieval Merchant Guilds of South India, Manohar Publications.

[88] Barnes, R. (1993), Indian Block-Printed Cotton Fragments in the Kelsey Museum, The University of Michigan Press, Ann Arbor.

[89] Christie, J. (1993), Texts and Textiles in Medieval Java, Bulletin de l'Ecole Francaise d'Extreme-Orient 80, 1, pp. 193–95.

[90] Christie, J. (1998), The medieval Tamil-language inscriptions in Southeast Asia and China, JSEAS, 29, 2, 257–8, Sastri, History, pp. 132–3.

a pastoral existence in the peripheral regions.[91] The staple food in Chola India was rice—and fermented rice soups—and it was traded for either barter or for gold, but also acted as a common measure of value by itself.[92] However, other commonly eaten groceries included dhal, curd, salt, spices such as cardamom, pepper, mustard, cumin, and betel leaves. Fish was also eaten alongside venison meat sold by forest hunters.[93] The latter suggests that salt and lemons were taxable commodities.

Trade and Plunder

At the end of the Sangam period, the three Tamil kingdoms disappeared for three centuries.[94] The new Chola dynasty emerged in around 850 AD, when Vijayalaya conquered Tanjavur and set up his capital there.[95] According to the latter, one of the greatest Indian kings came to the Chola throne in 985 AD, Rajaraja 1, who would rule until 1014 AD. In general, Chola warfare can be seen as an economic activity bent on profit as plunder rather than as profit as territorial gain.[96] In this context, according to the latter, the Chola kings used plunder to satisfy two strategies. Firstly, to increase a sustained supply of resources to the royal court, while plunder may be short term, it is different from long-term methods of financing the state through levies and tributes. The former was based on a quantification of land, goods, population and produce while the

[91] Subrahmanyam, S., and Bayly, C. (1988), Portfolio capitalists and the political economy of early modern India, The Indian Economic and Social History Review, 25, 4.

[92] Sakhuja, V., and Sakhuja, S. (2009), Rajendra Chola 1's Naval Expedition to Southeast Asia, in 'Nagapattinam to Suvarnadwipa: Reflections on the Chola Naval Expeditions to Southeast Asia', Kulke, H., Kesavapany, K., and Sakhuja, V. (Eds), Institute of Southeast Asian Studies, Singapore.

[93] Ibid.

[94] Kainikara, S. (2016), From Indus to Independence: A Trek Through Indian History, Volume II, The Classical Age, Vij Books India Pvt. Ltd., New Delhi.

[95] Heitzman, J. (1997), Gifts of Power: Lordship in an Early Indian State, Oxford University Press, Delhi.

[96] Spencer, G. (1976), The Politics of Plunder: The Cholas in Eleventh-Century Ceylon, The Journal of Asian Studies, Vol. 35, No. 3, pp. 405–419.

latter depended on the size of a local chieftain's military force.[97] And, secondly plundering involved the central co-ordination of a decentralised socio-political system. However, in the case of Rajaraja 1 (985 AD–1014 AD) and his son Rajendra 1 (1012 AD–1044 AD), this strategy evolved into plunder and the establishment of semi-permanent camps from which more plundering could be conducted.[98] This was the strategy followed by the Chola kings of the new dynasty under Rajaraja 1 and Rajendra 1 in Ceylon, present-day Sri Lanka, perhaps to gain control of the island due to its strategic location in the Arabian-China maritime trade link.[99] According to the latter, the Chola kings were engaged in regional and long distance trade through ports they controlled, with military support, perhaps in order to benefit from the levies which such trade would bring. Under the kingship of Rajendra 1, the Chola empire stretched from its Tamilakam heartlands to Karnataka, Andhra Pradesh, and Kedah in modern Malaysia.[100] Furthermore, the Chola influence also extended to Cambodia and Indonesia, with Tamil merchants recorded to have been in Sumatra.[101] Chola naval power was exerted in the Melaka Straits, in which the largely Malay population depended on maritime trade for their subsistence.[102] The Chola navy was used to defeat the Srivijaya Malay kingdom and bring it under the control of the Chola Empire in 1025 AD.[103] The mighty Chola Naval expedition of 1025 AD to Southeast Asia was unprecedented in Indian history.[104] The rise of the

---

[97] Ibid.

[98] Spencer, G. (1976), The Politics of Plunder: The Cholas in Eleventh-Century Ceylon, The Journal of Asian Studies, Vol. 35, No. 3, pp. 405–419.

[99] Ibid.

[100] Stein, B. (1977), Circulation and the Historical Geography of Tamil Country, The Journal of Asian Studies, Vol. 37, No. 1, pp. 7–26.

[101] Karashima, N., Subbarayalu, Y., and Shanmugam, P. (2008), Nagaram During the Chola and Pandyan Period: Commerce and Towns in Tamil Country AD 850–1350, The Indian Historical Review, Volume XXXV, No. 1, pp. 1–33.

[102] Heng, D. (2013), State formation and the evolution of naval strategies in the Melaka Straits, c.500–1500 CE, Journal of Southeast Asian Studies, 44(3), pp. 380–399.

[103] Kuiper, K. (2011), Understanding India: The Culture of India, Britannica Educational Publishing, New York.

[104] Kulke, H. (2009), The Naval Expeditions of the Cholas in the Context of Asian History, in 'Nagapattinam to Suvarnadwipa: Reflections on the Chola Naval Expeditions to Southeast Asia', Kulke, H., Kesavapany, K., and Sakhuja, V. (Eds), Institute of Southeast Asian Studies, Singapore.

Cholas in southern India in 985 AD synchronised with the rise of the Song dynasty in China in 960 AD and the Fatimids in Egypt in 969 AD. The rise of the Fatimids in Egypt coincided with the decline of the Abbasids dynasty in Baghdad. The shift in Middle Eastern power from Baghdad to Egypt caused a shift in the focus of Muslim trading activities from the Persian Gulf to the Red Sea. This caused the decline in trade with India's west coast, and increased trade with the Cholas and the Malabar coast.[105] Ships from the Red Sea ports were able to take advantage of the monsoon winds and sail quite easily to the South Indian Malabar coast. This shift increased Chola trade, which required a navy to defend and expand. The shift of dynasties and increased trade placed southern India at the centre of maritime trade between the Middle East and China. The result was that the Cholas would protect and expand their trading networks, eliminating competing powers like the Srivijaya kingdom in Sumatra. One reason put forward for the Chola military expansion in southern India and in Southeast Asia has been the need to acquire and stabilise sources of gold, precious stones, and wealth, so that the nobility and the military could continue to make donations to the ever-expanding Temples in southern India, including the Rajarajesvaram Temple.[106] Moreover, the Srivijaya kingdom was interfering with Chola trade with Song dynasty China.[107] This may have either been overtly or transparently, because of the perception by the Song court that the Chola state was weaker and subject to the Srivijaya kingdom, leading to Tamil merchants receiving fewer trading opportunities in China. Furthermore, in 987 AD, the Chinese government under the Song dynasty monopolised foreign trade by establishing the Bureau of Licensed Trade, which bought all imported goods and sold them at its own prices to the people.

[105] Abulughod, J. (1991), Before European Hegemony: The World System AD.1250–1350, Oxford University Press Inc., New York.

[106] Kulke, H., and Rothermund, D. (1998), A History of India, Routledge, New York.

[107] Sen, T. (2009), The Military Campaigns of Rajendra Chola and the Chola-Srivijaya-China Triangle, in 'Nagapattinam to Suvarnadwipa: Reflections on the Chola Naval Expeditions to Southeast Asia', Kulke, K., Kesavapany, K., and Sakhuja, V. (Eds), Institute of Southeast Asian Studies, Singapore.

So financially lucrative was this for the Song dynasty that they sent diplomatic missions to the Southeast Asian nations and the Cholas, encouraging them to increase trade with Song dynasty China. As an incentive, the Song dynasty government promised these trading nations that they would be given special facilities and import licences.[108] Economically, tax revenues from trade became more important for China following the An Lushan rebellion in 755 AD during the Tang dynasty, which, although it failed, caused the dynasty to relax an extremely rigid economic system in which China was essentially a closed economy, opening the economy up to foreign trade.[109] The dismantling of the rigid economic system also saw the emergence of private entrepreneurship.[110] Moreover, the planting of new types of crops from Southeast Asia and the development of irrigation techniques facilitated a surge in China's population from the late Tang dynasty and increased migration from China's interior region to its coastal areas, where the demand for foreign products mainly lay.[111] The incentivisation of trade by the Song dynasty in China ushered in a global economy in the tenth century, which consisted of the Middle East-South and Southeast Asia-China tripartite.[112] The linking of market trade with the tribute system by Song dynasty China also led to increased competition between foreign merchants, (Sen, 2009). The latter suggests that the greater the tribute presented to the Song court, by merchants on behalf of foreign kingdoms, the higher the tax rebates which the foreign merchants received. The Chola Emperor Rajaraja 1 sent the first Chola mission to the Song dynasty in 1015 AD, carrying tribute goods to the Song

---

[108] Sen, T. (1995), Maritime Contacts between China and the Chola Kingdom of South India: 850–1279', in 'Marines, Merchants and Oceans: Studies in Maritime History', Matthew, K. (Eds), Manohar, New Delhi.

[109] Sen, T. (2009), The Military Campaigns of Rajendra Chola and the Chola-Srivijaya-China Triangle, in 'Nagapattinam to Suvarnadwipa: Reflections on the Chola Naval Expeditions to Southeast Asia', Kulke, K., Kesavapany, K., and Sakhuja, V. (Eds), Institute of Southeast Asian Studies, Singapore.

[110] Sen, T. (2003), Buddhism, Diplomacy and Trade: The Realignment of Sino-Indian Relations, 600–1400, University of Hawaii Press, Honolulu.

[111] Sen, T. (2003), Buddhism, Diplomacy and Trade: The Realignment of Sino-Indian Relations, 600–1400, University of Hawaii Press, Honolulu.

[112] Modelski, G., and Thompson, W. (1996), leading Sectors and World Powers: The Coevolution of Global Economics and Politics, Columbia University Press.

Emperor of a robe, a cap, pearls, ivory, and incense along with a personal letter from the Chola Emperor to the Song Emperor.[113] The latter also suggests that the close contact between the Chinese empire and South India continued under the Yuan dynasty between 1271 AD and 1368 AD, with both sides exchanging annual diplomatic missions and establishing settlements in Quanzhou, China, and Nagapattinam in South India. A Siva Temple being built in Quanzhou and a three-storey Chinese Pagoda in Nagapattinam, which existed until 1867 AD. The Melaka Straits was a vital maritime trade link between India and China centred around the Buddhist Srivijaya kingdom which flourished in parts present-day Indonesia, Malaysia, Cambodia, and Vietnam. Chinese monks would stop off at Buddhist monasteries in the Srivijaya kingdom in order to learn Malay and Sanskrit before continuing on their journey to India. Between 1017 and 1025, the Chola navy raided Malay ports in the Melaka Straits as well as in the Gulf of Siam.[114] Though this naval engagement may have been brief, it did allow for the extension of Chola influence over the northern Melaka Straits for the next century.[115] However, it must be recognised that the Chola navy was not made up of warships on permanent station ready for combat, but on a fleet built up when necessary from merchant trading vessels.[116] Moreover, Chola seafaring vessels had neither a magnetic compass nor rudder, and had to rely on a knowledge of weather systems, prevailing winds, and heavenly bodies to navigate ships from the Malabar coast to ports in Southeast Asia, and on to China.[117] However, the latter suggests that they had instruments for locating stars, speed measurements, as well as a flat bronze plate for

[113] Kulke, H. (2009), The Naval Expeditions of the Cholas in the Context of Asian History, in 'Nagapattinam to Suvarnadwipa: Reflections on the Chola Naval Expeditions to Southeast Asia', Kulke, H., Kesavapany, K., and Sakhuja, V. (Eds), Institute of Southeast Asian Studies, Singapore.

[114] Heng, D. (2013), State formation and the evolution of naval strategies in the Melaka Straits, c.500–1500 CE, Journal of Southeast Asian Studies, 44(3), pp. 380–399.

[115] Heng, D. (2013), State formation and the evolution of naval strategies in the Melaka Straits, c.500–1500 CE, Journal of Southeast Asian Studies, 44(3), pp. 380–399.

[116] Sakhuja, V., and Sakhuja, S. (2009), Rajendra Chola 1's Naval Expedition to Southeast Asia, in 'Nagapattinam to Suvarnadwipa: Reflections on the Chola Naval Expeditions to Southeast Asia', Kulke, H., Kesavapany, K., and Sakhuja, V. (Eds), Institute of Southeast Asian Studies, Singapore.

[117] Ibid.

measuring water depth. Furthermore, pigeons were used to locate land and the human hand was used to measure the altitude of the stars. Voyages were made from Nagapattinam, the main port and the seat of governance of the Cholas. Lighthouses were maintained day and night along the coast in order to allow ships to navigate safely to port without hazard.[118] The latter suggests that knowledge of scurvy existed, such that ships were furnished with 'martabans' jars containing a mixture of lemons, mangoes, and salted ginger pepper condiments to be consumed by sailors on long voyages to provide them with the quantities of vitamin C necessary to prevent scurvy. The defining innovation of the Chola state was its extended use of naval power to conquer distant states.[119] The Cholas did have a professional army composed of regiments of cavalry, elephants to carry bowmen, infantry, and swordsmen.[120] The latter notes that, in 1178 AD, a Chinese scholar wrote that the Cholas had around 60,000 elephants in the army. At the time, it was noted that the Chola army was huge in number, perhaps with as many as 900,000 soldiers.[121] However, it has been suggested that the longevity and effectiveness of the Chola army, unlike the armies of Macedonia, Rome, and Greece, was denigrated by a lack of discipline because of the army's close links to South Indian society.[122] According to the latter, in this case, the Chola army reflected the lack of bonding in Indian society, as well as its divisions. The centre of Indian civilisation was not political, but rather, one based on religion and ritual.[123] The temple was at the heart of Indian socioeconomic life, and through this, the King attained power. The Chola influence was such that Tamil inscriptions have been found at sites on the Malay peninsula (Kedah), in northern Sumatra (Lobu Tua) and in eastern China

[118] Ibid.

[119] Barua, P. (2003), The State at War in South Asia, Board of Regents of the University of Nebraska.

[120] Sastri, K. (1935), The Cholas, The University of Madras.

[121] Subbarayalu, Y. (2009), A Note on the Navy of the Chola State, in 'Nagapattinam to Suvarnadwipa: Reflections on the Chola Naval Expeditions to Southeast Asia', Kulke, H., Kesavapany, K., and Sakhuja, V. (Eds), Institute of Southeast Asian Studies, Singapore.

[122] Barua, P. (2003), The State at War in South Asia, Board of Regents of the University of Nebraska.

[123] Barua, P. (2003), The State at War in South Asia, Board of Regents of the University of Nebraska.

(Quanzhou).[124] Nevertheless, by the second half of the twelfth century, the Srivijaya navy had gained the upper hand in the control of the Melaka Straits maritime trade due to the withdrawal of the Chola navy due to upheavals at home.[125] The latter suggests that this is in contrast to the first half of the twelfth century in which the Cholas were able to maintain diplomatic and economic relations with Song dynasty China. However, there are contradicting claims that direct Chola trade with China ceased in the latter part of the eleventh century, only for Chinese-South Indian trade to be resurrected by the Yuan dynasty in the latter part of the twelfth century.[126]

End of Empire

The Chola empire collapsed finally in the thirteenth century due to repeated and successful attacks from the Pandyans from the far south, the Hoysalas from Karnataka, and from local chieftains in Tamil Nadu itself.[127] The combined attacks of the Pandyans and the Hoysalas weakened the Chola state to such an extent that when King Rajendra IV, the last king of the Cholas, died in 1279 AD.[128] The lands of the Chola dynasty were then incorporated into the emergent Pandyan kingdom.[129] The rise of the military strength of the Pandyans and the Hoysalas, and the decreasing military strength of the Cholas can be explained by two

[124] Christie, J. (1998), The medieval Tamil-language inscriptions in Southeast Asia and China, JSEAS, 29, 2, 257–8, Sastri, History, pp. 132–3.

[125] Heng, D. (2013), State formation and the evolution of naval strategies in the Melaka Straits, c.500–1500 CE, Journal of Southeast Asian Studies, 44(3), pp. 380–399.

[126] Christie, J. (1998), The medieval Tamil-language inscriptions in Southeast Asia and China, JSEAS, 29, 2, 257–8, Sastri, History, pp. 132–3.

[127] Heitzman, J. (1987), Temple Urbanism in Medieval South India, Journal of Asian Studies, Vol46, No. 4, pp. 791–826.

[128] Mahalakshmi, R. (2016), Chola Empire, IN MacKenzie, J. (Ed), The Encyclopedia of Empire, John Wiley & Sons Ltd., London.

[129] Schmidt, K. (1995), An Atlas and Survey of South Asian History, Routledge, New York.

factors.[130] Firstly, the development of advanced forms of agriculture such as the use of flooded fields to grow rice seedlings;[131] according to the latter, this agricultural innovation occurred in China in the ninth century AD. This would have led to an increase in the yield of rice crops, which would facilitate an increase in the population as well as an agricultural surplus that could be used for either internal and external trade, or put in store for later use. The increase in the population led to an increase in the size of cities, and increased the possibilities for the 'division of labour'.[132] Secondly, peasant chiefs learned that their wealth could buy power. The increase in population discussed previously also facilitated an increase in the size of the ruling class.[133] The ideological innovations, and the favourable environment which gave rise to the ascendancy of the Cholas, were no longer valid.[134] Furthermore, the exchange and accessibility strategy the Cholas followed to maintain power were no longer effective.

The Chola navy continued to receive fewer and fewer resources and was not the fighting force it had been at the time of the conquest of the Srivijaya kingdom in 1025 AD, or at the time of support for the Chola Malay dependent states in 1078 AD. So, by the time the Chola Empire was brought to an end in 1261 AD, by the rise of the neighbouring Pandyan kingdom, the Chola navy had been in a state of hibernation.[135] The demise of the Chola state was also recognised by the Chinese writer Zhao Rugua, based in Quanzhou in 1225 AD, who also recognised that while the Chola empire was engaged in a fight for survival, it was still

[130] Stein, B. (1985), Politics, Peasants and the Deconstruction of Feudalism in Medieval India, in Feudalism and Non-European Societies, Byres, T., and Mukhia, H. (Eds), Frank Cass and Company Limited, London.

[131] Gough, K. (1980), Modes of Production in Southern India, Economic and Political Weekly, Vol. 15, No. 5/7.

[132] Ibid.

[133] Ibid.

[134] Stein, B. (1985), Politics, Peasants and the Deconstruction of Feudalism in Medieval India, in Feudalism and Non-European Societies, Byres, T., and Mukhia, H. (Eds), Frank Cass and Company Limited, London.

[135] Roy, K. (2015), Warfare in Pre-British India – 1500 BCE to 1740 BCE, Routledge, New York.

managing to export textiles.[136] Following the demise of the Chola state in the thirteenth century, the merchant guilds continued, but were in decline until the seventeenth century, functioning without the state support they had enjoyed under the umbrella of the Chola empire. However, as the merchant guilds declined individual traders on the South Indian coast carried on exporting textiles and dyes to the Middle East and Southeast Asia.[137] The fourteenth-century Vijayanagar kingdom incorporated the remnants of the Chola empire. Despite the demise of the Chola empire, it did leave its mark on southern India and Southeast Asia through its innovations in architecture, metallurgy, naval power, and 'political' ideology. Architectural innovation was evidenced in the design and construction of temples such as the Rajarajeswaram Temple at Thanjavur, which at the time of its completion was the tallest standing building in the world. The Cholas also made innovations in metallurgy with regards to the production of bronze statutes. This metallurgical innovation may have extended to the production of swords and other metallic fighting implements. However, the most prominent innovation of the medieval Chola dynasty was the scale of the use of naval power to subdue the other kingdoms on the Malay peninsula and in Sumatra and Java. This military expansion by the Cholas was to gain more trade with China to acquire wealth that could be donated to the temples in order to gain divine favour with the Hindu deity. The logistics and the coordination required for the transport of soldiers, elephants, and the required provisions on ships is astounding. This is especially in the context of the lack of modern communications we have today. The Cholas also innovated in the establishment of a political ideology through exchange and accessibility in order to stabilise and prolong the power of the King and the dynasty. However, the development and the growth of the agricultural sector allowed local chieftains to accumulate wealth, which was possible due to the way in which revenues from the proceeds of the land were distributed. Rajaraja had changed the pre-existing

[136] Hirth, F., and Rockhill, W. (1911), Chau Ju-Kua: His Work on the Chinese and Arab Trade in the Twelfth and Thirteenth Centuries, entitled Chu-fan-chi, Paragon Press, New York.

[137] Christie, J. (1998), The medieval Tamil-language inscriptions in Southeast Asia and China, JSEAS, 29, 2, 257–8, Sastri, History, pp. 132–3.

distribution system so that the state and the temples would receive a much bigger share. The 'state' may have included local chieftains. In conjunction with exchange and accessibility, the increasing wealth of the local chieftains may have given them enough confidence to usurp the power of the Chola king at a time when the Chola state itself was under attack and in a weak condition. Perhaps, by colluding with the enemies of the Cholas to strengthen their power.

References

Abraham, M. (1988), Two Medieval Merchant Guilds of South India, Manohar Publications.

Abulughod, J. (1991), Before European Hegemony: The World System AD.1250–1350, Oxford University Press Inc, New York.

Ball, W. (2016), Rome in the past: The Transformation of an Empire, 2nd Edition, Routledge, Oxon.

Barnes, R. (1993), Indian Block-Printed Cotton Fragments in the Kelsey Museum, The University of Michigan Press, Ann Arbor.

Barrett, D. (1964), An Early Cola Lingodbhavamurti, The British Museum Quarterly, Vol. 28, No. 1/2, pp. 32–39.

Barua, P. (2003), The State at War in South Asia, Board of Regents of the University of Nebraska.

Basham, A. (1954), The Wonder that was India, Sidgwick and Jackson, London.

Champakalakshmi, R. (1996), Trade, Ideology and Urbanization in South India 300 BC to AD 1300, Oxford University Press, Delhi.

Chaurasia, R. (2008), History of Ancient India: Earliest Times to 1200AD, Atlantic Publishers & Distributors (p) Ltd., New Delhi.

Christie, J. (1993), Texts and Textiles in Medieval Java, Bulletin de l'Ecole Francaise d'Extreme-Orient 80, 1, pp. 193–95.

Christie, J. (1998), The medieval Tamil-language inscriptions in Southeast Asia and China, JSEAS, 29, 2, 257–8, Sastri, History, pp. 132–3.

Cribb, J. (2003), The Origins of the Indian Coinage Tradition, South Asian Studies, 19, 1, pp. 1–19.

Danino, M. (1988), Vedic Roots of Early Tamil Culture, Vivekananda Kendra Patrika, Vol. 40, No. 1, 79th Issue.

DeRouen, K., and Heo, U. (2005), Defense and Security: A Compendium of National Armed Forces and Security Policies: Vol. 1, ABC-CLIO, Santa Barbara, CA.

Gough, K. (1980), Modes of Production in Southern India, Economic and Political Weekly, Vol. 15, No. 5/7.

Hall, K. (1980), Trade and Statecraft in the Age of the Colas, Abhinav Publications, New Delhi.

Heitzman, J. (1987), Temple Urbanism in Medieval South India, Journal of Asian Studies, Vol. 46, No. 4, pp. 791–826.

Heitzman, J. (1991), Ritual Polity and Economy: The Transactional Network of an Imperial temple in Medieval South India, Journal of the Economic and the Social History of the Orient, Vol. 34, No. ½, pp. 23–54.

Heitman, J. (1997), Gifts of Power: Lordship in an Early Indian State, Oxford University Press, Delhi.

Heng, D. (2013), State formation and the evolution of naval strategies in the Melaka Straits, c.500–1500 CE, Journal of Southeast Asian Studies, 44(3), pp. 380–399.

Hirth, F., and Rockhill, W. (1911), Chau Ju-Kua: His Work on the Chinese and Arab Trade in the Twelfth and Thirteenth Centuries, entitled Chu-fan-chi, Paragon Press, New York.

Howard, M. (2012), Transnationalism in Ancient and Medieval Societies: The Role of Cross-Border Trade and Travel, McFarland & Company, North Carolina.

Irwin, J. (1948), South Indian Figure Sculpture, 7th–10th Centuries AD, Journal of the Royal Asiatic Society.

Kainikara, S. (2016), From Indus to Independence: A Trek Through Indian History, Volume II, The Classical Age, Vij Books India Pvt Ltd., New Delhi.

Karashima, N. (1984), South Indian History and Society: Studies from Inscriptions, AD 850–1800, OUP, New Delhi.

Karashima, N. (1996), South Indian Temple Inscriptions: A New Approach to their study, South Asia: Journal of Asian Studies, 19,1, 1–12.

Karashima, N. (2008), Temple land in Chola and Pandyan inscriptions: The legal meaning and historical implications of Kudiningadevadana, Indian Economic and Social History Review. 45, 175–99.

Karashima, N., and Subbarayalu, Y. (2004), The Emergence of the Periyanadu Assembly in South India during the Chola and Pandyan Periods, International Journal of Asian Studies, 1,1, pp. 87–103.

Karashima, N., Subbarayalu, Y., and Shanmugam, P. (2008), Nagaram During the Chola and Pandyan Period: Commerce and Towns in Tamil Country AD 850–1350, The Indian Historical Review, Volume XXXV, No. 1, pp. 1–33.

Krishnan, K.G. (1982), ChittiramemlMip-periyanamdDu – An Agricultural Guild of Medieval Tamil Nadu. Journal of the Madras University, Vol. 54:1

Kuiper, K. (2011), Understanding India: The Culture of India, Britannica Educational Publishing, New York.

Kulke, H. (2009), The Naval Expeditions of the Cholas in the Context of Asian History, in 'Nagapattinam to Suvarnadwipa: Reflections on the Chola Naval Expeditions to Southeast Asia,' Kulke, H., Kesavapany, K., and Sakhuja, V. (Eds), Institute of Southeast Asian Studies, Singapore.

Kulke, H., and Rothermund, D. (1998), A History of India, Routledge, New York.

Lippe, A. (1960), The Sculpture of Greater India, The Metropolitan Museum of Art Bulletin, New Series, Vol. 18, No. 6, pp. 177–192.

Mahalakshmi, R. (2016), Chola Empire, In MacKenzie, J. (Ed), The Encyclopedia of Empire, John Wiley & Sons Ltd., London.

Mendis, G. (2005), The Early History of Ceylon: And Its Relations With India and Other Foreign Countries, Asian Educational Services, New Delhi, Chennai.

Modelski, G., and Thompson, W. (1996), Leading Sectors and World Powers: The Coevolution of Global Economics and Politics, Columbia University Press.

Mukund, K. (2012), The World of the Tamil Merchant: Pioneers of International Trade, Penguin Books, Haryana, India.

Prichard, P. (1995), Tanjavur Brhadisvara: An Architectural Study, Indira Gandhi Centre for the Arts, New Delhi.

Reddy, V. (2009), Temples of South India, Guyan Publishing House, New Delhi.

Roy, K. (2015), Warfare in Pre-British India – 1500BCE to 1740BCE, Routledge, New York.

Sastri, K. (1935), The Colas, The University of Madras.

Sastri, K. (1958), A History of South India from Pre-historic Times to the fall of Vijayanagar, Oxford University Press, London.

Sakhuja, V., and Sakhuja, S. (2009), Rajendra Chola 1's Naval Expedition to Southeast Asia, in 'Nagapattinam to Suvarnadwipa: Reflections on the Chola Naval Expeditions to Southeast Asia,' Kulke, H., Kesavapany, K., and Sakhuja, V. (Eds), Institute of Southeast Asian Studies, Singapore.

Schmidt, K. (1995), An Atlas and Survey of South Asian History, Routledge, New York.
Sen, T. (1995), Maritime Contacts between China and the Chola Kingdom of South India: 850–1279, in 'Marines, Merchants and Oceans: Studies in Maritime History,' Matthew, K. (Eds), Manohar, New Delhi.
Sen, T. (2003), Buddhism, Diplomacy and Trade: The Realignment of Sino-Indian Relations, 600–1400, University of Hawaii Press, Honolulu.
Sen, T. (2009), The Military Campaigns of Rajendra Chola and the Chola-Srivijaya-China Triangle, in 'Nagapattinam to Suvarnadwipa: Reflections on the Chola Naval Expeditions to Southeast Asia,' Kulke, K., Kesavapany, K., and Sakhuja, V. (Eds), Institute of Southeast Asian Studies, Singapore.
Sen, T. (2010), The Military Campaigns of Rajendra Chola and the Chola-Sri Vijaya-China Triangle. In H. Kulke et al. (Eds.), Nagapattinam to Suvarnadwipa: Reflections on the Chola Naval Expeditions to Southeast Asia: 61–75.
Spencer, G. (1976), The Politics of Plunder: The Cholas in Eleventh-Century Ceylon, The Journal of Asian Studies, Vol. 35, No. 3, pp. 405–419.
Srinivasan, S. (1998), The use of tin and bronze in prehistoric southern Indian metallurgy, *Journal of Metals*, 50(7), 44–9.
Stein, B. (1977), Circulation and the Historical Geography of Tamil Country, The Journal of Asian Studies, Vol. 37, No. 1, pp. 7–26.
Stein, B. (1985), Politics, Peasants and the Deconstruction of Feudalism in Medieval India, in Feudalism and Non-European Societies, Byres, T., and Mukhia, H. (Eds), Frank Cass and Company Limited, London.
Subbarayalu, Y. (2009), A Note on the Navy of the Chola State, in 'Nagapattinam to Suvarnadwipa: Reflections on the Chola Naval Expeditions to Southeast Asia,' Kulke, H., Kesavapany, K., and Sakhuja, V. (Eds), Institute of Southeast Asian Studies, Singapore.
Subrahmanyam, S., and Bayly, C. (1988), Portfolio capitalists and the political economy of early modern India, The Indian Economic and Social History Review, 25, 4.
Subrahmanyam, S. (1990), The Political Economy of Commerce: Southern India, 1500–1650, Cambridge University Press, Cambridge.
Thapar, R. (1990), A History of India, Vol. 1, Penguin Books, London.
Talwai, P. (2010), Praying through Politics, Ruling Through Religion: The Rajarajeswaram as an instrument of Economic and Political Unification in the Chola Empire, UC Berkeley, scholarship, https://escholarship.org/uc/item/70v1v9dh

Vasudevan, G.(2003), The Royal Temple of Rajaraja: An Instrument of Imperial Chola Power. New Delhi: Abhinav.
Venkataraman, B. (1985), Rajarajesvaram: The Pinnacle of Chola Art, Mudgala Trust, Madras.
Venkatachari, K.K.A. (1987), Rajaraja the Great: Seminar Proceedings, Ananthacharya Indological Research Institute, Bombay.
Wilkinson, D. (1996), Configurations of the Indic States System, Comparative Civilisations Review, 34.

8

The Han Dynasty (206 BC to 220 AD) and the Song Dynasty (960 AD to 1279 AD)

Introduction

The Qin dynasty collapsed fifteen years after it had unified China.[1] The Qin dynasty, though repressive, had offered China peace and stability after 500 years of war and instability.[2] However, after the collapse of the Qin dynasty, there followed further civil strife in which the military forces of Liu Bang and Xiang Yu fought each other for the control of the lands of the Qin. The kings of the Liu family were to form the Han dynasty.[3] In order to stop five years of civil war, the nobles and the generals gathered on the banks of the River Fan on 28 February, 202 BC, in order to proclaim Liu Bang as the first Emperor of the Han dynasty, which would last for over four hundred years.[4] Liu Bang was born a peasant and showed

[1] Feng, Li. (2013), Early China: A Social and Cultural History, Cambridge University Press, New York.

[2] Hardy, G., and Kinney, A. (2005), The Establishment of the Han Empire and Imperial China, Greenwood Press, London.

[3] Rawson, J. (1999), The Eternal Palaces of the Western Han: A New View of the Universe, Artibus Asiae, Vol. 59, No. 1/2, pp. 5–58.

[4] Hardy, G., and Kinney, A. (2005), The Establishment of the Han Empire and Imperial China, Greenwood Press, London.

S. Ramesh, *The Rise of Empires*, https://doi.org/10.1007/978-3-030-01608-1_8

humility in refusing to become Emperor at the urging of his advisors, following the death of his rival Xiang Yu. However, Liu Bang finally accepted the title in order to bring peace and stability to China.[5] The latter suggests that institutions implemented by the Han dynasty were continued by successive dynasties until the collapse of the last of China's dynasties, the Qing dynasty, in 1912. The Han dynasty was matched in terms of military power, territorial and population size; and cultural strength, only by the Roman Empire.[6] Geographically, the Han dynasty can be represented as the shift between the highlands of the upper Yellow River and the flood plains of the lower Yellow River.[7] The Song dynasty, on the other hand, was formed several hundred years later and existed in the tenth, eleventh, and twelfth centuries AD. It was formed in southern China as a result of losing the northern Chinese heartlands to marauding northern tribes such as the Khitan. The southern Song dynasty was the heartbeat of innovation in economic systems, navigation and maritime trade, and a naval military force.

The Han Dynasty (206 BC–220 AD)

Early Governance

The governance structures put in place by the Qin dynasty in 221 BC survived in its basic form for 2000 years, until the end of dynastic rule in China in 1911.[8] However, the governance structures were amended in a number of ways, firstly, by the Han dynasty, which overthrew the Qin. The main difference between the Han dynasty and the Qin dynasty was that the rule of the former was based on the mandate of heaven while the latter relied on a model of political change based on earth, wood, metal,

---

[5] Ibid.

[6] Ibid.

[7] Lewis, M. (2009), China Between Empires: The Northern and Southern Dynasties, The Belknap Press of Harvard University Press, Cambridge, Massachusetts.

[8] Yu-ch'uan, W. (1949), An Outline of The Central Government of the Former Han Dynasty, Harvard Journal of Asiatic Studies, Vol. 12, No. 1 / 2, pp. 134–187.

fire, and water.[9] According to the latter, the first Emperor of the Qin dynasty defeated the previous Zhou dynasty through the use of overpowering military force, and not by claiming the mandate of heaven. This refers to the divine right to rule by a family given to it by heaven, which could be revoked if the common people suffered.

The commanderies (provinces) and the kingdoms were the two types of regional governments which existed under the Han dynasty.[10] The commanderies themselves were formed of a number of counties, the counties administered themselves, but the counties in a commandery were administered at the commandery level. The county was the third lowest form of administrative region in both the commanderies as well as in the kingdoms. Commanderies had also been established in what is modern-day Korea at the time of the Emperor Wu of Han (141 BC–87 BC).[11] The counties were divided into districts and the districts into 'li'.[12] Several commanderies made up a kingdom. At the start of the Han period, each kingdom was made up of at least six commanderies, and the kingdoms made up at least two-thirds of the country, with the remaining third of the country being made up of regional commanderies.[13] The latter suggests that the role of king in the kingdoms was hereditary in nature, and the King was responsible for appointing administrators at both the commandery and county levels.[14] Nevertheless, although the role of king was hereditary, the Emperor could ultimately replace a royal house and appoint a king of his choosing. This is what Liu Bang did as the first emperor of the Han dynasty, appointing several members of his family as kings in several kingdoms. This was in the belief that familial links would

[9] Hardy, G., and Kinney, A. (2005), The Establishment of the Han Empire and Imperial China, Greenwood Press, London.

[10] Edwards, R. (2009), Federalism and the Balance of Power: China's Han and Tang Dynasties and the Roman Empire, Pacific Economic Review, 14:1, pp. 1–21.

[11] Pai, H. (1992), Culture contact and culture change: The Korean peninsula and its relations with the Han Dynasty commandery of Lelang, World Archaeology, 23:3, 306–319.

[12] Yu-ch'uan, W. (1949), An Outline of The Central Government of the Former Han Dynasty, Harvard Journal of Asiatic Studies, Vol. 12, No. 1 / 2, pp. 134–187.

[13] Edwards, R. (2009), Federalism and the Balance of Power: China's Han and Tang Dynasties and the Roman Empire, Pacific Economic Review, 14:1, pp. 1–21.

[14] Bielenstein, H. (1980), The Bureaucracy of Han Times, Cambridge University Press, Cambridge.

breed loyalty. Liu Bang replaced all the kings in place at the time of his ascendancy to Emperor with family members, except in Changsha.[15] According to the latter, when Liu Bang died in 195 AD, the control of the imperial throne was taken by Empress Lu, one of his consorts, who appointed two puppet emperors and put to death the sons of other concubines who had become kings under Liu Bang's rule; and replaced by Lu family members. In some cases, land was also taken away from the kingdoms ruled by Liu kings in order to supplement the kingdoms of the newly appointed Lu kings.[16] However, the ascendancy of the Lu kings was brief.

Institutional Reforms

Some of the earliest institutional reforms of the Han date back to the first emperor, Hui Tsung, and his Grand Councillor, Wang An-Shih, who was in office in the period 1070 AD to 1076 AD, at different times. The state monopolisation of commercial activity, the equalisation of prices by the state, the organisation of local militia for the defence of the country; and the substitution of the conscription of labour, for example. For when the Empress Lu died in 180 AD, a son of Liu Bang, the Emperor Wendi, ascended to the throne. As a result, Lu kings were dethroned, but a mistrust between the imperial court and the regional kings continued to exist.[17] In order to reduce the possibility of rebellion by the kings, the Emperor Wendi reduced the size of existing kingdoms and increased the number of kingdoms and commanderies.[18] The stability of this strategy relied on the existence of a central unifying authority, the Emperor, and a ruling dynasty. During successive periods of China's history, when this

---

[15] Edwards, R. (2009), Federalism and the Balance of Power: China's Han and Tang Dynasties and the Roman Empire, Pacific Economic Review, 14:1, pp. 1–21.

[16] Ibid.

[17] Ibid.

[18] Loewe, M. (1986), The Former Han Dynasty, in Twitchett, D., and Loewe, M. (Eds), The Cambridge History of China – Volume 1, The Ch'in and Han Empires, 221 BC–220 AD, Cambridge University Press, Cambridge.

was not possible, China ceased to exist as a single entity. For example, during the period of the three kingdoms—220 AD to 280 AD—and the period of the five dynasties and ten kingdoms 907 AD to 960 AD. However, the increase in the number of kingdoms continued to fuel the tensions between the imperial court and the regional kings. As a result, more institutional reforms were instigated when the Emperor Wendi died in 157 BC, and was replaced by his son, the Emperor Jingdi. These institutional changes can be mainly associated with the imperial court taking direct control over the resource-rich commanderies of some of the most powerful kingdoms.[19] This proved to be destabilising, as the kings resented the eroding of their power. As a result, the Seven Kings Rebellion against the imperial court took place in 154 BC. This rebellion was suppressed by the imperial court with the despatch of two imperial armies.[20] Rebellious kings were killed and replaced. However, the institutional changes made by the Emperor Jingdi were more far reaching because the kings lost their power to appoint administrators and officials within their domains. Many posts were eliminated, specifically, those with fiscal and military responsibilities.[21] The Kings roles became titular in nature, and they were 'pensioned' off, receiving an income from the revenues raised in their kingdoms.[22] The Emperor Jingdi was succeeded by his son, the Emperor Wudi, in 141 BC. Under the Emperor Wudi, the centralising institutional reforms were continued.[23] According to the latter, these institutional changes continued with the reduction in the size of kingdoms through either the establishment of new kingdoms or the formation of new commanderies. Agriculture was also well developed in the Han dynasty, with at least five grains in production, rice—the principal

[19] Ibid.

[20] Joyner, H. (1978), The Recruitment, Organisation, and Control of Former Han Military Leadership, Unpublished Master's Thesis, Department of Far Eastern Languages and Civilisations, University of Chicago.

[21] Edwards, R. (2009), Federalism and the Balance of Power: China's Han and Tang Dynasties and the Roman Empire, Pacific Economic Review, 14:1, pp. 1–21.

[22] Bielenstein, H. (1980), The Bureaucracy of Han Times, Cambridge University Press, Cambridge.

[23] Edwards, R. (2009), Federalism and the Balance of Power: China's Han and Tang Dynasties and the Roman Empire, Pacific Economic Review, 14:1, pp. 1–21.

crop—millet, hemp, wheat, and soybeans.[24] Irrigation and fertiliser was also used. The agricultural techniques used during the Han dynasty would be used to boost production in the Song dynasty, during which high yielding strains of agriculture were experimented with and put into production.[25] However, Chang Ch'ien, sent on a diplomatic mission to West Asia by the Emperor Wudi, brought back new crops such as the grapevine, alfalfa, soybeans, and horses.[26] The system of accounting inherited from the Qin dynasty was improved upon. The invention of paper and the abacus during the latter stages of the Eastern Han Dynasty (25 AD to 220 AD) facilitated the better record keeping of transactions as well as an improved method of calculation.[27] The Han government adopted an accounting system that depended upon the determination of government expenditures and revenues. The latter arose from taxes, of which there were four main types.[28] There were taxes on merchants, artisans, the land, and on males between the ages of 15 and 56. Under the Han dynasty, tax levies were increased because taxes were the main source of government revenues.[29] Moreover, according to the latter, during the Han dynasty, the Shangji system of reporting was introduced. It was an annual conference during which the financial accounts of various regions would be presented to the Vice Premier, the Inspector of Taxes, and the Emperor himself. The Emperor would be able to ask questions. In this way, officials in regions doing well could be rewarded whilst officials in regions doing badly could be punished and measures needed to improve performance introduced.

---

[24] Ballas, D. (1965), Some Notes on Agriculture in Han China, The Professional Geographer, 17:4, pp. 13–14.

[25] Golas, P. (1980), Rural China in the Song, Journal of Asian Studies, Vol. XXXIX, No. 2.

[26] Hirth, F. (1917), The Story of Chang Ch'ien, China's Pioneer in West Asia, Journal of the American Oriental Society, Vol. 37, pp. 89–152.

[27] Aiken, M., and Lu, W. (1993), Historical Instances of Innovative Accounting Practices in the Chinese Dynasties and beyond, The Accounting Historians Journal, Vol. 20, No. 2, pp. 163–186.

[28] Chou, C. (1974), An Economic History of China, Western Washington State College Bellingham, Washington.

[29] Aiken, M., and Lu, W. (1993), Historical Instances of Innovative Accounting Practices in the Chinese Dynasties and beyond, The Accounting Historians Journal, Vol. 20, No. 2, pp. 163–186.

Between 9 AD and 23 AD, there was an interlude in the Han dynasty, with the usurpation of the Han throne by Wang Mang to form the Xin dynasty. Emperor Wang Mang proposed reforms which were akin to today's Socialist/Communist manifesto. Policies he championed included taking big holdings of land and dividing it up amongst the peasants.[30] However, drought and starvation fermented rebellion amongst the peasants, and his reforms fermented rebellion amongst the nobility. Wang Mang was killed in 23 AD, and the Xin dynasty came to an end with the beginning of the Eastern Han dynasty (25 AD to 220 AD).

Centralisation of Power

By the last decades of the second century AD, the Han empire included the territories comprising modern-day China, southern Manchuria, northern Korea, and Tonkin, but it did not include the extreme southwest.[31] The Han had relied on a belief in marriage diplomacy, marrying Han princesses to the leaders of confrontational tribes. However, under the Emperor Wu of Han (141 BC to 87 BC), the Han Empire's borders extended into Xinjiang, Yunnan, Korea, and Vietnam.[32] The latter suggests that, at the time, the Han empire counted 12.2 million households and 60 million individuals. It was also at the height of its military and economic power. And for the first time in China's history, the Han had introduced a centralised imperial government, with the Emperor possessing total authority over all his subjects. The governance of such a large population required a large and structured administrative network. This network was a hierarchy of twenty ranks, although, after 32 BC, this was

[30] Chih-P'ing, C. (2013), English Writings of Hu Shih, Chinese Philosophy and Intellectual History (Volume 2), Springer, London.

[31] Yu-ch'uan, W. (1949), An Outline of The Central Government of the Former Han Dynasty, Harvard Journal of Asiatic Studies, Vol. 12, No. 1 / 2, pp. 134–187.

[32] Chin, T. (2010), Defamiliarizing the Foreigner: Sima Qian's Ethnography and Han-Xiongnu Marriage Diplomacy, Harvard Journal of Asiatic Studies, Vol. 70, No. 2, pp. 311–354.

reduced to sixteen ranks.[33] The latter suggests that the rank assigned to a bureaucrat would indicate his status, salary, privileges, what type of clothes he wore, and what type of carriage he could ride. At the head of this administrative hierarchy was the Emperor. The Emperor appointed officials with a salary of six hundred bushels. These officials included those in the central government cabinet and, at a local level, the Emperor also appointed provincial governors, county prefects, and county chiefs. As well as being the source of all laws, the Emperor was also the ultimate judge over all matters. Moreover, while in pre-Han China, being in government and having political power also meant having wealth and social prestige. However, post-Han, the Emperor was the only source of power.[34] However, in two cases, the enthronement of Emperor Wendi (179 BC to 157 BC) and Emperor Hsuan (73 BC to 49 BC) was decided by a debate between nobles and ministers in a court conference.[35]

The assistant to the Emperor was the Chancellor, who was responsible for all the officials in the service of the Emperor as well as state finances. In addition to the Chancellor, there were also the roles of Tutor and Guardian. The Chancellor, the Tutor, and the Guardian were known as the 'Three Councillors', instigated in the reign of the Chou, abolished by the Qin but reinstated by the Han dynasty.[36] The latter suggests that there were also two standing armies, one in the capital and a national army. The capital army was controlled by military officers under the direction of three governors. The national army was commanded by a general and four lieutenant generals, known as the right, left, front, and the rear.[37] The Supreme Court was instigated by the Qin dynasty but survived into the Han dynasty, being attached to the Palace, with the Chief Justice being appointed by the Emperor.[38] The ministers of importance in the

[33] Yu-ch'uan, W. (1949), An Outline of The Central Government of the Former Han Dynasty, Harvard Journal of Asiatic Studies, Vol. 12, No. 1 / 2, pp. 134–187.

[34] Ibid.

[35] Ibid.

[36] Koo, T. (1920), The Constitutional Development of the Western Han Dynasty, Journal of the American Oriental Society, Vol. 40, pp. 170–193.

[37] Ibid.

[38] Ibid

central government were the Three Lords (San Kung) and the Nine Ministers (Chiu Ch'ing).[39] Other important government positions included the Commandant of Justice and the Grand Herald. The former settled legal disputes which could not be resolved by the provincial governors, while the latter was responsible for the integration into Han society of barbarians who had submitted to the rule of the Emperor. Between the period 194 BC and 141 BC, the Han Empire was ruled by a 'Politik' of the children of the founding father of the dynasty, Liu Bang, whose friends had also become ministers.[40] The latter suggests that, by 140 BC, the traditional Minister in the imperial cabinet had been replaced by Confucian scholars. Knowledge and learning had a very prominent place in ancient China, especially during the Han dynasty. However, it may have been the case that three schools of thought—Confucianism, Legalism, and Taoism—may have been prevalent in the imperial court, with officials having leanings towards particular schools of thought.[41] The latter suggests that the intellectual environment in the Han period was also attached to the Way of the Tao of Taoism. Some of the assertions of Taoism related to a belief in the specialisation of production, that people should specialise in the production of a good or service to which their innate skills was proficient.[42] At the same time, there should also be a complementary evolution of the tradespeople in general in order to facilitate exchange.[43] The scholarly tradition continued in succeeding dynasties. Knowledge and learning related to a process of accumulation, dissemination, and refinement.[44] In Mathematics, the Southern Song mathematician Qin Jiushao laid out the principles for solving high order algebraic expressions in the 'Shu-shu Chiu-Chang', also known as the

[39] Yu-ch'uan, W. (1949), An Outline of The Central Government of the Former Han Dynasty, Harvard Journal of Asiatic Studies, Vol. 12, No. 1 / 2, pp. 134–187.

[40] Ibid.

[41] Kaufman, C. (1987), The evaluation of marketing in a society: The Han dynasty of ancient China. Journal of Macromarketing 7:52–64

[42] Kaufman, C. (1987), The evaluation of marketing in a society: The Han dynasty of ancient China. Journal of Macromarketing 7:52–64

[43] Ibid.

[44] Baark, E. (2007), Knowledge and Innovation in China: Historical Legacies and Emerging Institutions, Asia Pacific Business Review, Vol. 13, No. 3, pp. 337–356.

'Nine Chapters of Mathematics' in 1247 AD. However, the work derived much benefit from the mathematical techniques developed in the Han period.[45] The work of Qin Jiushao related to using mathematics to solve real-world problems such as surveying, taxation, construction, as well as the relationship between interest and price. In astronomy, there were also developments, with Liu Hong devising the first complete Chinese theory of the moon.[46] And in literature, there was a focus on history and the works of Pan Ko and Ssuma Ch'ien, the 'History of the Former Han Dynasty' and the 'Records of the Historian' respectively, emerged.[47] This literary tradition continued into the fourteenth century, with the publication of the 'Romance of the Three Kingdoms'. This book depicted the events in the late Han dynasty and the machinations of the Emperor Cao Cao.[48]

Chinese civilisation has always had a reverence for learning, and this has been recognised as a cornerstone of traditional Chinese culture.[49] This was derived from the Confucian ideal that learning cultivated intellectually and morally able men. Scholars were placed at the highest level in the Confucian scale of values, to be followed by farmers, artisans, and merchants.[50] According to the latter, this form of social organisation was very popular in Han China. However, at a wider level, the scholars, farmers, artisans, and merchants were the commoners, and were positioned at the penultimate level of the Han social ladder, just above the slaves. The Emperor and the royal family, the nobles and the officials preceded the

[45] Wang, L., Ling, W., and Needham, J. (1955), Horner's Method in Chinese Mathematics: Its Origins in the Root Extraction Procedures of the Han Dynasty, T'oung Pao, Second Series, Vol. 43, Livr. 5, pp. 345–401.

[46] Cullen, C. (2002), The First Complete Chinese Theory of the Moon: The Innovations of Liu Hong, Journal for the History of Astronomy, 33;1–24.

[47] Kaufman, C. (1987), The evaluation of marketing in a society: The Han dynasty of ancient China. Journal of Macromarketing 7:52–64

[48] Wang, L., Ling, W., and Needham, J. (1955), Horner's Method in Chinese Mathematics: Its Origins in the Root Extraction Procedures of the Han Dynasty, T'oung Pao, Second Series, Vol. 43, Livr. 5, pp. 345–401.

[49] Baark, E. (2007), Knowledge and Innovation in China: Historical Legacies and Emerging Institutions, Asia Pacific Business Review, Vol. 13, No. 3, pp. 337–356.

[50] Ibid.

commoners in that order.[51] Confucianism became more prominent in the Han period, as there was a requirement that potential government employees should do well in the civil service entrance exams. However, while Confucianism was the state adopted school of thought, there were also other critical thinkers, especially during the period of the Eastern Han. These thinkers included Wang Chong, Wang Fu, Xu Gan, and Xun Yue.[52] Wang Chong and Xu Gan were more innovative thinkers than Wang Fu and Xun Yue, both of whose work revolved around Confucianism. Wang Chong wrote critically on a varying range of subjects, which included physics, anatomy, cosmology, ethics, and meteorology.[53] According to the latter, Xu Gan developed the concept of 'actuality', or the state of development, which led to the name of an entity and how this may have been shattered.

Any knowledge (scientific, technical, and cultural) which was generated could be easily disseminated because of the ready availability of printing technology. The focus of analysis was the order of the cosmos and the Dao, behaviour.[54] The latter also suggests the nature of the analysis was, in the context of the universe, encompassing heaven, earth, society, and the individual. However, although China was innovative, inventing paper, printing, gunpowder, and the magnetic compass. Paper was invented by Cai Lun during the time of the Eastern Han Dynasty, and printing was invented in 220 AD.[55] However, while gunpowder was produced in the ninth century by unknown alchemists, its original purpose was medicinal and not for warfare. The innovation of the compass did not take place until after the Han dynasty, during the Song period. These innovations did not have as far reaching consequences for China's

---

[51] Kaufman, C. (1987), The evaluation of marketing in a society: The Han dynasty of ancient China. Journal of Macromarketing 7:52–64

[52] McLeod, A. (2015), Philosophy in Eastern Han Dynasty China (25 – 220 CE), Philosophy Compass, 10/6, pp. 355–368.

[53] Ibid.

[54] Baark, E. (2007), Knowledge and Innovation in China: Historical Legacies and Emerging Institutions, Asia Pacific Business Review, Vol. 13, No. 3, pp. 337–356.

[55] Van Someren, T.C.R. & Van Someren-Wang, S. (2013), Chinese culture, strategy, and innovation. Innovative China: Innovation Race between East and West (pp. 2756). Berlin: Springer

economic development as it did for Europe.[56] Perhaps the Chinese were not as creative in devising applications for these innovations, as were the Europeans due to a lack of political and economic stability. Maybe the swing to one dynasty after another over China's long history ensured that there was never any cultural, economic, or political cohesion to allow for the further innovation of accumulated knowledge. Or perhaps, the Confucian ideology of 'knowing one's place in society' limited creativity and abstract science.[57]

According to Confucian theory, the best form of governance was that by benevolent rulers. In other words, the fate of the empire depends on the upright behaviour of the Emperor.[58] During the reign of the Emperor Wudi, there was much conflict between the Emperor and the Imperial Cabinet. This necessitated the creation of a new position to act as a go-between between the Emperor and the Cabinet, the Grand Minister of the Mount, or the Prime Minister.[59] The Prime Minister acted as a regent on behalf of the Emperor, with unlimited power, but subordinate to the Chancellor. However, the impact on the imperial government of creating the regency was that the positions of Chancellor and Imperial Secretary were relegated to that of mere administrators. Moreover, the Chancellor had been 'replaced' by the someone close to the Emperor, the Regent, and the government was now referred to as the Inner Court rather than the Outer Court, because government now moved to the Palace.[60] The latter suggests that another signature feature of the transfer of political power from day-to-day ministers to individuals close to the Emperor was the appointment of eunuchs and Masters of Documents. The latter were lowly officials in the service of the bursar of the imperial household. The promotion of the Masters of Documents occurred because they had

[56] Shaffer, L. (1986), China, Technology and Change, World History Bulletin 4, No. 1, pp. 1–6.

[57] Tsou, C. (1998), Science and Scientists in China, Science, New Series, Vol280, No. 5363, pp. 528–529.

[58] Yu-ch'uan, W. (1949), An Outline of The Central Government of the Former Han Dynasty, Harvard Journal of Asiatic Studies, Vol. 12, No. 1 / 2, pp. 134–187.

[59] Koo, T. (1920), The Constitutional Development of the Western Han Dynasty, Journal of the American Oriental Society, Vol. 40, pp. 170–193.

[60] Yu-ch'uan, W. (1949), An Outline of The Central Government of the Former Han Dynasty, Harvard Journal of Asiatic Studies, Vol. 12, No. 1 / 2, pp. 134–187.

gained the trust of the Emperor through frequent political consultations that had occurred because of their preparation of decrees and rescripts.[61] According to the latter, the Masters of Documents would keep notes on officials throughout the empire, so they could give information to the Emperor on those officials most meriting promotion.

Impacts of Centralised Power

Although the Han Emperor did not 'own' the empire, he did have two valuable resources.[62] These included the labour of the people and the taxes which could be collected. These taxes included a land tax and a poll tax on adults (Suan-fu) and children (K'ou-ch'ien).[63] According to the latter, further taxes on merchant's property and on livestock were introduced in 119 BC and 114 BC respectively. The revenues from the taxes allowed the Emperor to pay his administrative officials and maintain a standing army that could be called upon to defend the borders of the Empire as well as to maintain internal control. The ranks of the army were also increased by conscription. Individuals were required to serve two years in the army: one year for training and the second year for stationing at a garrison.[64] Moreover, according to the latter, individuals were also required to carry out annual monthly military service in their localities. Furthermore, the central government gaining strength over time allowed it to control the production and supply of iron and salt as well as coinage in 119 BC and 115 BC respectively.[65] Nevertheless, to unify the people and win their loyalty, the Emperor also embarked on the building of ancestral temples in his empire. This has some resonance with the activities of the Chola dynasty in tenth-century southern India. However,

[61] Ibid.

[62] Yu-ch'uan, W. (1949), An Outline of The Central Government of the Former Han Dynasty, Harvard Journal of Asiatic Studies, Vol. 12, No. 1 / 2, pp. 134–187.

[63] Ibid.

[64] Ibid.

[65] Ibid.

by the middle of the first century BC, the Han empire was much weaker due to a combination of socioeconomic factors.[66]

Agriculture also saw development during the Han period. In this case, two modes of production co-existed: the peasant holding and the 'manor'—in which there was one landlord and many peasant farmers.[67] The latter suggests that, in the early Han period, the peasant farmer holding was the basic unit of agricultural production, with government policy focusing on eradicating the manor or the large estate. The reason for this was that manor land was farmed by peasant farmers on behalf of the land owner. The land owner was supposed to take the tax from the peasants and pass it on to the government. However, the land owner kept a portion of the tax for themselves and charged the peasant farmers a rent. The state was losing income, and the only way this could be stopped was by eradicating manors and large estates. The small peasant holdings were integrated in to the wider market economy but, at unstable times, the peasant farmer returned to subsistence living, but these periods in the 400 years of the Han period were relatively short. The government provided the infrastructure necessary for the peasant farmers to increase production and interventionist measures to allow the peasant farmers to maintain their independence.[68] The infrastructure projects included the building of canals in Honan and Shansi in order to irrigate six million acres of arable land. Other small-scale irrigation projects were also undertaken in the Huai and the Wei River valleys as well as in Northwest China.[69] The latter suggests that, as a direct result of the irrigation projects, the crop yield in the lower Yellow River Valley was six million bushels of grain by the middle part of Emperor Wu-Ti's reign (140 BC to 87 BC). However, the crop yield had only been a few hundred thousand bushels before the irrigation works. The government also

[66] Ibid.

[67] Bray, F. (1979), Agricultural Technology and Agrarian Change in Han China, Early China, Vol. 5, pp. 3–13

[68] Bray, F. (1979), Agricultural Technology and Agrarian Change in Han China, Early China, Vol. 5, pp. 3–13

[69] Ibid.

provided the seeds and the agricultural tools to the peasant farmers as loans or as gifts. To increase agricultural production, year-round production was encouraged in different parts of the country. Moreover, to grow crops during winter time, experiments were undertaken to grow crops in pits in which they could be protected by frost and cold weather to some extent. However, in the latter Han period, the rise of the large estates or manors followed the rise of large powerful families. The large estates were able to rationalise agricultural production in order to suit the needs of commercial production.[70] In a modern economics context, commercial production and commercial activity requires formalised contracts. However, in Han literature, there was no mention of contracts as one would think in a modern context.[71] The latter notes that this inattention was despite the widespread commercial activity and the Han tradition of enforcing contracts. However, in a cultural context, there was the concept of 'Yueh', or binding. The latter is a Confucian notion that knowledge allows an individual to know what is proper, and so, can refrain from doing anything improper. In other words, at a cultural and a philosophical level, there is a moral obligation for individuals to honour an agreement, although such an agreement is not represented in a formalised contract. Another ethical quality attached to the formation of informal contracts was that of 'hsin', which can be equivalently translated as trustworthiness or credibility.[72] In the Han period, the mediation of disputes was undertaken in the belief that there had been some form of negligence on the part of the mediator, the scholar. If the people in dispute can be taught well, then there would never have been a dispute in the first place. This is a central Confucian belief.[73] Whereas Roman Law was split into criminal and civil law, and then to include

[70] Ibid.

[71] Scogin, H. (1990), Between Heaven and Man: Contract and the State in Han Dynasty China, 63 S. Cal. L. Rev. 1325.

[72] Scogin, H. (1990), Between Heaven and Man: Contract and the State in Han Dynasty China, 63 S. Cal. L. Rev. 1325.

[73] Pei, C. (1999), The Origins of Mediation in Traditional China, Dispute Resolution, 54, 2.

canon law, Chinese law was not, and the civil law strand was replaced by Confucian ethics.[74] But this approach takes human nature at its face value and does not account for the cunning and Machiavellian behaviour human beings actually exhibit in practice. Therefore, legal protection of inventions, in the form of patents, represent the path to economic efficiency and prosperity.[75] This is clearly evidenced if a comparison is made between the legal protection available for invention in seventeenth/eighteenth-century England and China, whether Han, the Song, or any other Chinese dynasty before 1911. In the case of England, patents gave inventors the incentives, income, to invent. On the other hand, in China, because of the lack of patent protection, income for inventors was not sure so potential inventors studied for the civil service exams in order to secure a well payed and secure government job.[76] But this did not mean that there was no innovation in Han China. For example, in the early Han period, the best-known patterned bronze mirrors in China began to appear.[77]

The law in China came to be a symbol of imperial tyranny and power ever since the Qin dynasty tried to eliminate Confucian ethics.[78] The penal code in Han China was issued as statutes, ordinances, or regulations.[79] In the context of state power, statutes were the most important. This is the role that the law took in China from the time of the Han dynasty.[80] However, the symbolism for the political legitimacy of imperial

[74] Lehman, J. (2006), Intellectual Property Rights and Chinese Tradition Section: Philosophical Foundations, Journal of Business Ethics, 69:1–9.

[75] Li, R. (2009), Private Property Rights, Legal Enforcement and Economic Prosperity: The Fall of Early Civilised China and the Rise of the United Kingdom in the 18th–19th century Asian Science Journal, Vol. 5, No. 10.

[76] Ibid.

[77] Cammann, S. (1948), The 'TLV' Pattern on Cosmic Mirrors of the Han Dynasty, Journal of the American Oriental Society, Vol. 68, No. 4, pp. 159–167.

[78] Lehman, J. (2006), Intellectual Property Rights and Chinese Tradition Section: Philosophical Foundations, Journal of Business Ethics, 69:1–9.

[79] MacCormack, G. (2004), The Transmission of Penal Law (Lü) from the Han to the T'ang: A Contribution to the Study of the Early History of Codification in China, Revue Internationale des Droits de l'Antiquité LI (2004): 47–83.

[80] Lehman, J. (2006), Intellectual Property Rights and Chinese Tradition Section: Philosophical Foundations, Journal of Business Ethics, 69:1–9.

regimes in China arose from the name of the state, dynasty, and the associated cyclical cosmic power.[81]

In the latter Han period, big investments were needed in terms of seeds and tools, which the small peasant holder farmer was unable to provide. Moreover, a growing population meant that small land holdings were becoming less feasible. According to the latter, the main regions of agricultural production in Han China encompassed the lands surrounding the lower part of the Yellow River and the valley of the Wei River. This region of China had been farmed continuously over a period of 5000 years, since Neolithic times. The soil of the valley of the Wei River was fertile, but more susceptible to water loss through evaporation. On the other hand, the soils of the land surrounding the Yellow River, were much 'heavier', perhaps due to its ability to retain moisture. In these lands, drainage was the main problem.[82] In the case, of the valley of the Wei River, a light plough was used, with ploughing done at a time when the soil held enough moisture. The plough did not penetrate the soil by more than 3–4 inches. On the other hand, the lands surrounding the lower Yellow River required the use of the mouldboard plough. The use of the mouldboard plough has numerous advantages. These advantages included the use of less seed, which could be thinned allowing the resulting plants to be spaced correctly, and the convenience of hoeing.[83]

As the Han dynasty collapsed and centralised power became decentralised through the three kingdoms, the Jin, and the Northern and Southern dynasties, a sense of individualism in 'social, philosophical and literary ideas' emerged.[84] According to the latter, this is best evidenced by the quantity and the quality of biographical works that appeared after the end of the Han dynasty.

---

[81] Chan, H. (2008), The 'Song' Dynasty Legacy: Symbolism and Legitimation From Han Liner to Zhu Yuanzhang of the Ming Dynasty, Harvard Journal of Asiatic Studies, Vol. 68, No. 1, pp. 91–133.

[82] Bray, F. (1979), Agricultural Technology and Agrarian Change in Han China, Early China, Vol. 5, pp. 3–13.

[83] Ibid.

[84] Chen, S. H. (1953). An Innovation in Chinese Biographical Writing. The Journal of Asian Studies, 13, 49–62.

The Song Dynasty (960 AD to 1279 AD)

The Northern Song Dynasty (960 AD to 1127 AD)

By 907 AD, the Tang dynasty had been brought to an end having been weakened by the An Lushan rebellion in 755 AD.[85] Between 907 AD and 959 AD, China disintegrated and was not united again until 960 AD, by the founder of the Song dynasty (960 AD to 1279 AD), the beginnings of which was the Northern Song Dynasty (960 AD–1126 AD). In spite of this, northern China remained under the control of the Jin dynasty, the forest dwelling Jurchen people of northern Manchuria, who were also skilled horsemen.[86] However, the Song were able to move the capital to the south of the country, to Kaifeng in modern-day Henan province, where the Emperor presided over the central government.[87] At the time, there were six ministries: Revenue, Civil Appointments, Rites, Works, Punishment, and War. The heads of these six ministries reported to the Grand Councillor, who was appointed by the Emperor. Unlike the Han dynasty, where the government had been moved to the palace, the Song emperors were able to either contribute to the functioning of government or to remain indifferent.[88] In this case, according to the latter, some Song emperors chose to delegate governance matters to the Grand Councillor and spend time in leisurely and scholarly activity, while other emperors chose to show a direct interest in the governance of their empire. The latter suggests that government officials were selected on the basis of performance in the civil service exams, which was one of the innovations of the Song dynasty. The transition from the Tang dynasty to the Song dynasty left in place the political structures that had existed since the time of the Han dynasty. The prefectures had

[85] Hansen, V., and Curtis, K. (2010), Voyages in World History, Volume 1, Wadsworth, Cengage Learning, Boston, USA.

[86] Ibid.

[87] Ibid.

[88] Ibid.

been the largest administrative features in the Tang dynasty, and in the Song dynasty there were 220 prefectures that were grouped into 20 larger structures called circuits.[89] The latter suggests that each prefecture in a circuit was further divided into sub-prefectures that were governed by a magistrate and his staff, their main role being to administer the Emperor's justice and to collect imperial taxes. However, one important change made by the first Song emperor was to transfer control of regions from the military to bureaucrats.[90] This was due to the perception that the Tang dynasty had crumbled because of too much power in the hands of regional governors who had been generals. However, the Northern Song dynasty was brought to an end by the Jin dynasty, following an alliance between them to defeat the nomadic Khitan tribe who inhabited what would be recognised as present-day Mongolia.[91] The Jin cavalry proved to be swift and effective in defeating the Song army, and Kaifeng was besieged and captured in late 1127 AD with the use of explosive devices made using bamboo, gunpowder, and porcelain shards.[92] The Jin would continue to rule over 40 million Chinese until 1234.[93] The capture of Kaifeng indicated the wonders of Chinese architecture to the non-Chinese nomadic northern tribes, such as the Khitan or the Liao to whom Chinese culture spread.[94] In addition to architectural knowledge, knowledge of calligraphy also spread. Both men and women practised calligraphy in the imperial court.[95] It is possible that knowledge of this aspect of Chinese culture may also have spread to the northern tribes such as the Khitan.

---

[89] Ibid.

[90] Hansen, V., and Curtis, K. (2010), Voyages in World History, Volume 1, Wadsworth, Cengage Learning, Boston, USA.

[91] Ibid.

[92] Ibid.

[93] Ibid.

[94] Kuhn, D. (2000), Liao Architecture: Qidan Innovations and Han-Chinese Traditions? T'and Pao, Second Series, Vol. 86, Fasc.4/5, pp. 325–362.

[95] Lee, H. (2004), The Emperors Lady Ghost-writers in Song-Dynasty China, Artibus Asiae, Vol. 64, No. 1, pp. 61–101.

The Southern Song Dynasty (1127 AD to 1279 AD)

One of the sons of the Emperor Huizong, the Northern Song emperor who reigned from 1101 AD to 1125 AD, escaped Kaifeng and became the first emperor of the Southern Song Dynasty in 1127 AD. This emperor established his capital in Hangzhou, in present-day Zhejiang province, and ruled over 60 million Chinese.[96] Those who fled Kaifeng and crossed the Yangzi River, following its conquest by the Jin included not only the first Southern Song Emperor but also government officials, clerks, soldiers, and the poet Li Qingzhao. It has been said that the fall of Kaifeng to the Jin caused one of the world's first biggest mass migration in history, in which over 500,000 people out of a total Chinese population of 100 million moved from the north to the south.[97] According to the latter, after years of warfare, the Jin and the Southern Song signed a peace treaty in 1142, according to which the Jin would be paid an annual tribute of silver and silk. All the tribute was paid by a strong state to a weaker state, and proved to be in the Southern Song's favour, as the Jin would use the tribute to buy Song goods. However, the loss of profitable lands in the North meant that, in the short term, the Song look to other sectors of the economy to make money. In this case, the Emperor Gao Zong (1127 AD to 1162 AD) viewed maritime trade, if managed properly, as a source of considerable profits.[98] Products produced in Song dynasty China, such as ceramics, were exported as far as the Philippines and Indonesia.[99] The latter suggests that, in this case, under the Emperor's direction, the government invested in improving facilities at southern coast ports, widening canals, building warehouses, and navigation

[96] Hansen, V., and Curtis, K. (2010), Voyages in World History, Volume 1, Wadsworth, Cengage Learning, Boston, USA.

[97] Hansen, V., and Curtis, K. (2010), Voyages in World History, Volume 1, Wadsworth, Cengage Learning, Boston, USA.

[98] Levathes, L. (1994), When China Ruled the Seas: The Treasure Fleet of the Dragon Throne, 1405–33, Open Road Distribution, New York.

[99] Little, S. (1996), Economic Change in Seventeenth-Century China and Innovations at the Jingdezhen Kilns, Ars Orientalis, Vol. 26, pp. 47–54.

beacons along the southern coast. Moreover, by 1132 AD, the navy was established as part of the structure of the Song military in order to protect trade routes and merchant ships from pirates. By 1237 AD, the Song navy grew into a professional fighting force that controlled the seas from China's southern ports to as far as Korea and Japan.[100] At the time, it also had the largest standing army in the world.[101] According to the latter, by 1237 AD, the Song navy comprised 52,000 men, 20 squadrons, and 600 vessels. Further innovations in ship design, ships weaponry, and navigation systems such as the maritime compass occurred at this time at the instigation of the Emperor.[102] Different parts of ships would be made from different woods. For example, China fir would be used to make the sternpost and planks, the frames from Camphor and pine, and cypress would be used to make the bulkheads.[103] When the ships planks and superstructure had been nailed together, the whole ship was waterproofed using a mixture of putty and jute fibres.[104] Furthermore, Chinese scholars innovated by studying the maritime traditions of other cultures, in particular, the Hindus and the Arabs.[105] However, by the early thirteenth century, command of ships was transferred from professional naval officers to the imperial court. Scholars and historians consider the period encompassing the Southern Song Dynasty as a 'commercial revolution'. The Song was the first dynasty in China to rely on commercial activity for tax revenues, rather than the agricultural sector.[106] To encourage commercial activity, the Song government either did not raise taxes, or it

[100] Levathes, L. (1994), When China Ruled the Seas: The Treasure Fleet of the Dragon Throne, 1405–33, Open Road Distribution, New York.

[101] Golas, P. (1980), Rural China in the Song, Journal of Asian Studies, Vol. XXXIX, No. 2

[102] Levathes, L. (1994), When China Ruled the Seas: The Treasure Fleet of the Dragon Throne, 1405–33, Open Road Distribution, New York.

[103] Kenderdine, S. (1995), Bai Jiao 1 – the Excavation of a Song Dynasty Shipwreck in the Dinghai Area, Fujian Province, The International Journal of Nautical Archaeology, 24, 4, pp. 247–266.

[104] Guo-Qing, L. (1989), Archaeological evidence for the use of 'Chu-nam' on the 13th Century Quanzhou Ship, Fujian Province, China, The International Journal of Nautical Archaeology and Underwater Excavation, 18, 4, pp. 277–283.

[105] Levathes, L. (1994), When China Ruled the Seas: The Treasure Fleet of the Dragon Throne, 1405–33, Open Road Distribution, New York.

[106] Grass, N. (2017), The Economic History of China – From Antiquity to the Nineteenth Century, Ming Studies, 2017:76, 109–112.

chose to lower it.[107] Merchants kept accounts to determine whether they were making a profit or a loss from their business activities. A four-column reporting system, developed in the Tang dynasty, was used and elaborated upon throughout the Song period.[108] According to the latter, the equation identified with this system was (old + newly received= payment + balance).[109] The government also published annually a book on national accounts and accounts.[110] Due to the level of commercial activity and economic growth, there were more and more financial accounts to record and balance. As a result, the accounting system during the Song dynasty became more sophisticated than at any other time in traditional Chinese history.[111] Nevertheless, developments in accounting systems were less at the start of the Song dynasty. But the pace of development increased as commercialism took hold.[112] The latter suggests that increasing commercial activity brought credit sales into fashion. In conjunction with the issuing of paper money by the Song government, there was a need for accounting systems to change in order to incorporate these developments.

There was no political ideology which could explain this detachment of tax revenues from the land.[113] However, unlike previous dynasties, the Song were more reliant on market forces to resolve social welfare. Nevertheless, beginning during the Song period, the concept of the

[107] Aiken, M., and Lu, W. (1993), Historical Instances of Innovative Accounting Practices in the Chinese Dynasties and beyond, The Accounting Historians Journal, Vol. 20, No. 2, pp. 163–186.

[108] Ibid.

[109] Lin, J. (1992), Chinese Double-entry Book-keeping Before the Nineteenth Century, The Accounting Historians Journal, The Academy of Accounting Historians, Vol. 19, No. 2, pp. 103–122.

[110] Aiken, M., and Lu, W. (1993), Historical Instances of Innovative Accounting Practices in the Chinese Dynasties and beyond, The Accounting Historians Journal, Vol. 20, No. 2, pp. 163–186.

[111] Fu, Y. (1968), A Study of Governmental Accounting in China: with Special Reference to the Song Dynasty, PhD Dissertation, University of Illinois at Urban-Champaign.

[112] Lin, J. (1992), Chinese Double-entry Book-keeping Before the Nineteenth Century, The Accounting Historians Journal, The Academy of Accounting Historians, Vol. 19, No. 2, pp. 103–122.

[113] Grass, N. (2017), The Economic History of China – From Antiquity to the Nineteenth Century, Ming Studies, 2017:76, 109–112.

'clan' became very important in Chinese society.[114] The clan provided the welfare support, which the Song state never did.[115] The latter suggests that the clan was also responsible for tax collection, dispute resolution, the conduct of its members, as well as the preparation of clan members for the civil service entrance exams. At feasts, traditional Chinese folk instruments such as the 'Zheng' would be played according to clan custom.[116]

The clans acted as an enforcement mechanism for the state. It, therefore, did not maintain a granary of surplus grain for the 'hard' times, but instead depended on the seasons to balance the shortfalls of grain from one year to the next.[117] The commercial revolution in Song China, emerged due to a change in the local ecosystem and innovation in the agricultural sector of the south of China. Unlike the north of China, the southern lands were swampy and mosquito ridden, and this limited the availability of land for the growing of crops. However, the new migrants from the war-ridden north drained the swamps, walled fields, and built agricultural channels. Farmers experimented with different varieties of rice and found that a strain of rice from Vietnam had a short lifecycle, allowing for two annual harvests of rice.[118] The Southern Song, therefore, experienced a surplus in rice production. This had many implications for Song society. Firstly, farmers' incomes increased, and they could buy other goods. This in turn created a market for handicrafts and allowed the agricultural sector to diversify into other types of agricultural production, such as the production of fruits and textiles. As demand for textiles grew, agricultural workers left the land to produce textile fibre from hemp or

[114] Fei, J., and Ts'ui-Jung, L. (1982), The Growth and Decline of Chinese Family Clans, Journal of Interdisciplinary History, 12(3): 375–408

[115] Greif, A., and Tabellini, G. (2010), Cultural and Institutional Bifurcation: China and Europe Compared, The American Economic Review, Vol. 100, No. 2, pp. 135–140.

[116] Zheng, C., and Knobloch, Y. (1983), A Discussion of the History of the Gu Zheng, Asian Music, Vol. 14, No. 2, Chinese Music History, pp. 1–16.

[117] Grass, N. (2017), The Economic History of China – From Antiquity to the Nineteenth Century, Ming Studies, 2017:76, 109–112.

[118] Hansen, V., and Curtis, K. (2010), Voyages in World History, Volume 1, Wadsworth, Cengage Learning, Boston, USA.

silk, or a combination of both.[119] These textile workers became more innovative over time, developing new types of weaves of increasing complexity. Furthermore, small but effective technological innovations increased the productivity of the textile workers.[120] Iron production also increased, and China's output of iron, in the twelfth century, was not matched by any other country until England experienced its industrial revolution in the late eighteenth century. Secondly, the increased production of rice meant that the population of the south experienced dramatic increases. Such was the case that, while in 742 AD 60% of the population lived in the north, by 980 AD, 62% of the population lived in the south of China.[121] The increase in population also meant that the size of urban centres, cities, and towns also began to increase. Allied to increased trade and commercial activity, the rise of urban centres also meant that the number of disease and illness causing pathogens was also on the rise.[122] Under the direction of the Emperor and the government, there was a revival of interest in medicine. This exhibited itself in three ways.[123] Firstly, there was an increase in the number of medical books published and distributed. Secondly, there was a revival of interest in traditional medicines, practices and doctrines—things which had fallen by the way side since the demise of the Han dynasty. Thirdly, the number of medicines used to treat illnesses doubled during the period of the Northern Song dynasty. Finally, in the Song dynasty, the constituents of medicines as well as the dosage prescribed changed, both of which, before the Song dynasty, had remained unchanged for over 1000 years.

As the handicrafts and the market for other goods expanded, the Song authorities increased the money supply. People were no longer buying goods just for subsistence, **(Eckhardt et al. 2010).** The development of

---

[119] Sheng, A. (1990), Textile Use, Technology and Change in Rural Textile Production in Song China (960AD to 1279AD), PhD Dissertation, University of Pennsylvania.

[120] Ibid.

[121] Hansen, V., and Curtis, K. (2010), Voyages in World History, Volume 1, Wadsworth, Cengage Learning, Boston, USA.

[122] Goldschmidt, A. (2005), The Song Discontinuity: Rapid Innovation in Northern Song Dynasty Medicine, Asian Medicine – Tradition and Modernity, 1.1:53–90.

[123] Ibid.

printing allowed the money supply to be composed of not only coins, but also paper money, which entered circulation in 1000 AD.[124] The paper money entered circulation in Song China at a time when the technology was diffusing to the Middle East, but had not yet reached Europe. In 1069 AD, a new Grand Councillor instigated policy which called for government officials to be paid their salaries using paper money.[125] But, under pressure to increase military expenditure, in order to protect the land from central Asian invaders, the Song government printed too much paper money, which subsequently lost its value due to rising prices.[126] At this time, a new policy that facilitated the availability of low interest loans for peasants was also launched as an initiative of the Grand Councillor. However, the peasants were unable to repay the loans and, over time, this policy failed. Rising incomes and the expansion of the market also allowed innovations to be effective. For example, before 700 AD, book printing was a laborious process but, after 700 AD, the use of the woodblock allowed books to be printed much more quickly and in large numbers.[127] Rising incomes and the lower cost of books allowed reading to take off, children could be educated more effectively, and more individuals could sit the civil service exams.[128] Books on topics such as medicine, astronomy, agriculture, and geography were published in significant numbers and widely distributed to the population.[129] The poet, philosopher, and statesman, Wang Anshi (1021 AD to 1086 AD) wrote the 'Wan Yan Shu'.[130] This was a text which related to how the civil service should be

[124] Hansen, V., and Curtis, K. (2010), Voyages in World History, Volume 1, Wadsworth, Cengage Learning, Boston, USA.

[125] Ibid.

[126] Rowe, D. (2005), The Origins of Value, Goetzmann, W., and Rouwenhorst, G. (Eds), Global Association of Risk Professionals, Issue 6, pp. 28.

[127] Hansen, V., and Curtis, K. (2010), Voyages in World History, Volume 1, Wadsworth, Cengage Learning, Boston, USA.

[128] Ibid.

[129] Goodrich, L. C. (1963), The Development of printing in China and its effects on the renaissance under the Sung Dynasty (960–1279 CE), The Hong Kong Branch of the Royal Asiatic Society, 3: 36–43.

[130] Drechsler, W. (2013), Wang Anshi and the origins of modern public management in Song Dynasty China, Public Money & Management, 33:5, pp. 353–360.

reformed. Wang Anshi also proposed other reforms, including ones to strengthen the military. However, his ideas were resented by court officials because of the modernity of the reforms. Court officials wanted to enjoy the quiet and peaceful life the Chinese people had enjoyed in ancient times. In 1040 AD, a book printer called Bi Sheng invented a movable printing technology in which individual characters could be placed in a frame to print the required page.[131] Rising incomes and cheaper book production caused a boom in the educational sector.[132] The latter suggests that there was intense competition for government jobs. As a result, people who had studied for but failed the civil service exams were hired as tutors for those seeking to take the exams. The exams took place every three years in two rounds: once in the home prefecture and once in the capital. Although, the exams had been used for 1000 years, it was only in the Song dynasty that the majority of those in government service—90%—had taken and passed the exams. The implication is that the Song dynasty was the first in the history of China in which government officials were selected on the basis of birth rather than on their status.[133] According to the latter, the imperial civil service exams were designed to test a range of abilities. Candidates had to correctly identify texts from literary works of the Confucian tradition, write poems and essays of a fixed word count, and write essays on matters which should concern the government.[134] Academic learning was centred on learning and reproducing the same five classic texts; this may have acted as a constraint to creativity and innovative thinking.[135] However, the Song dynasty represents a period in which there was a rapid transformation of Chinese society in the context of demographic changes, changes in governance, learning, and innovation in association with a commercial revolution. This has led some scholars to suggest that, in the eleventh century, Song dynasty China was on the cusp of an industrial revolution—which did

---

[131] Hansen, V., and Curtis, K. (2010), Voyages in World History, Volume 1, Wadsworth, Cengage Learning, Boston, USA.

[132] Ibid.

[133] Ibid.

[134] Ibid.

[135] Van Someren, T.C.R. & Van Someren-Wang, S. (2013), Chinese culture, strategy, and innovation. Innovative China: Innovation Race between East and West (pp. 2756). Berlin: Springer.

not occur for another 500 years, until late eighteenth-century England.[136] However, the diversion of intellectual attention and capacity towards passing the civil service examinations may have prevented China's intellectuals from transitioning from primary research to secondary modern research.[137]

In ancient China, death rituals were an important moment for the immediate family members of the deceased to fulfil their filial duties.[138] This was one of the core values of the time. Death rituals were also symbolic of the expression of religious and philosophical belief. In eleventh century AD China, the prominent religious and philosophical leanings were towards either Buddhism or Daoism.[139] The values associated with both were expressed especially in the practice of death rituals, which were a way in which the wellbeing of the soul for the next life could be achieved. However, the extant way in which death rituals were conducted in ancient China also exhibited the economic disparities within that society.[140] For example, the grandness of the death ritual would depend upon the materials the dead would wear, as well as the size of the burial plot and the quality of the coffin. The authenticity of the death rituals also depended upon the way in which the classical texts were interpreted. These classical texts gave detailed instructions on how the corpse was to be washed, how it would be dressed, as well as how the burial was to be conducted.[141] In this context, there was a divergence between how officials and the imperial court interpreted the classical texts for the purposes of the performance of the death ritual.[142] The latter suggests that this was reflected by the

[136] Elvin, M. (1973), The Pattern of the Chinese Past: A Social and Economic Interpretation, Stanford University Press, Stanford.

[137] Lin, Y. J. (1995), The Needham Puzzle: Why the Industrial Revolution Did 'Not Originate in China', Economic Development and Cultural Change, Vol. 43, No. 2, pp. 269–292.

[138] Choi, M. (2017), Death Rituals and Politics in Northern Song China, Oxford University Press, New York.

[139] Ibid.

[140] Ibid.

[141] Standaert, N. (2008), The Interweaving of Rituals, Funerals in the Cultural Exchange between China and Europe, University of Washington Press, Seattle.

[142] Choi, M. (2017), Death Rituals and Politics in Northern Song China, Oxford University Press, New York.

charges brought against the officials by the imperial court for performing death rituals according to an inaccurate interpretation of the classical texts. Because of the conflict between officials and the imperial court, the performance of the death rituals came under state control with violations subject to the penal code. In eleventh-century Song dynasty China, there was, therefore, a convergence of the forces relating to filial family duties, the values of society, and state policies.[143] The conflict relating to the interpretation of the classical texts led to a revival of Confucianism in eleventh century AD Northern Song China, and the emergence of Neo-Confucianism or the Way of Learning Confucianism, Daoism, in twelfth century AD.[144] It was the scholars of Daoism, led by leading Song scholars such as Zhu Xi (1130 AD to 1200 AD), who studied the classical Confucian texts, and dissipated his findings to a wider society so that commoners would not engage in errant ritual burial practices.[145] In addition to a revival of Confucianism in the Song dynasty, there was also a continuation of Tang dynasty ritual practices such as the Taiqing Palace ritual and the Nine Palace ritual.[146] The former related to the celebration of the birthday of the sage Lao Zi, the founder of Taoism. The Nine Palaces ritual relates to the movement of energy through the universe, where energy is represented as the Pa-k'ua (wind, fire, mother, lake, father, water, mountain, thunder, and wind).[147]

References

Aiken, M., and Lu, W. (1993), Historical Instances of Innovative Accounting Practices in the Chinese Dynasties and beyond, The Accounting Historians Journal, Vol. 20, No. 2, pp. 163–186.

---

[143] Ibid.

[144] Ibid.

[145] Standaert, N. (2008), The Interweaving of Rituals, Funerals in the Cultural Exchange between China and Europe, University of Washington Press, Seattle.

[146] Xiong, V. (1996), Innovations and Taoism under Tang Xuanzong, T'oung Pao, Second Series, Vol. 82, Fasc. 4/5, pp. 258–316.

[147] Wong, E. (2011), Taoism: An Essential Guide, Shambhala Publications Inc., Boston, Massachusetts.

Baark, E. (2007), Knowledge and Innovation in China: Historical Legacies and Emerging Institutions, Asia Pacific Business Review, Vol. 13, No. 3, pp. 337–356.

Ballas, D. (1965), Some Notes on Agriculture in Han China, The Professional Geographer, 17:4, pp. 13–14.

Bielenstein, H. (1980), The Bureaucracy of Han Times, Cambridge University Press, Cambridge.

Bray, F. (1979), Agricultural Technology and Agrarian Change in Han China, Early China, Vol. 5, pp. 3–13.

Cammann, S. (1948), The 'TLV' Pattern on Cosmic Mirrors of the Han Dynasty, Journal of the American Oriental Society, Vol. 68, No. 4, pp. 159–167.

Chan, H. (2008), The 'Song' Dynasty Legacy: Symbolism and Legitimation From Han Liner to Zhu Yuanzhang of the Ming Dynasty, Harvard Journal of Asiatic Studies, Vol. 68, No. 1, pp. 91–133.

Chen, S. H. (1953). An Innovation in Chinese Biographical Writing. The Journal of Asian Studies, 13, 49–62.

Chih-P'ing, C. (2013), English Writings of Hu Shih, Chinese Philosophy and Intellectual History (Volume 2), Springer, London.

Chin, T. (2010), Defamiliarizing the Foreigner: Sima Qian's Ethnography and Han-Xiongnu Marriage Diplomacy, Harvard Journal of Asiatic Studies, Vol. 70, No. 2, pp. 311–354.

Choi, M. (2017), Death Rituals and Politics in Northern Song China, Oxford University Press, New York.

Chou, C. (1974), An Economic History of China, Western Washington State College Bellingham, Washington.

Cullen, C. (2002), The First Complete Chinese Theory of the Moon: The Innovations of Liu Hong, Journal for the History of Astronomy, 33;1–24.

Drechsler, W. (2013), Wang Anshi and the origins of modern public management in Song Dynasty China, Public Money & Management, 33:5, pp. 353–360.

Eckhardt, G., and Bengtsson, A. (2010), A Brief History of Branding in China, Journal of Macromarketing, 30(3), pp. 210–221.

Edwards, R. (2009), Federalism and the Balance of Power: China's Han and Tang Dynasties and the Roman Empire, Pacific Economic Review, 14:1, pp. 1–21.

Elvin, M. (1973), The Pattern of the Chinese Past: A Social and Economic Interpretation, Stanford University Press, Stanford.

Feng, Li. (2013), Early China: A Social and Cultural History, Cambridge University Press, New York.

Fei, J., and Ts'ui-Jung, L. (1982), The Growth and Decline of Chinese Family Clans, Journal of Interdisciplinary History, 12(3): 375–408

Ferguson, XX. ( ), Political Parties of the Northern Song Dynasty, http://e-asia.uoregon.edu.

Fu, Y. (1968), A Study of Governmental Accounting in China: with Special Reference to the Song Dynasty, PhD Dissertation, University of Illinois at Urban-Champaign.

Golas, P. (1980), Rural China in the Song, Journal of Asian Studies, Vol. XXXIX, No. 2.

Goldschmidt, A. (2005), The Song Discontinuity: Rapid Innovation in Northern Song Dynasty Medicine, Asian Medicine – Tradition and Modernity, 1.1:53–90.

Goodrich, L. C. (1963), The Development of printing in China and its effects on the renaissance under the Sung Dynasty (960–1279 CE), The Hong Kong Branch of the Royal Asiatic Society, 3: 36–43.

Grass, N. (2017), The Economic History of China – From Antiquity to the Nineteenth Century, Ming Studies, 2017:76: 109–112.

Greif, A., and Tabellini, G. (2010), Cultural and Institutional Bifurcation: China and Europe Compared, The American Economic Review, Vol. 100, No. 2, pp. 135–140.

Guo, D. (1988), The Historical Contributions of Chinese Accounting, Collected papers of the Fifth World Congress of Accounting Historians, The University of Sydney.

Guo-Qing, L. (1989), Archaeological evidence for the use of 'Chu-nam' on the 13th Century Quanzhou Ship, Fujian Province, China, The International Journal of Nautical Archaeology and Underwater Excavation, 18, 4, pp. 277–283.

Hansen, V., and Curtis, K. (2010), Voyages in World History, Volume 1, Wadsworth, Cengage Learning, Boston, USA.

Hardy, G., and Kinney, A. (2005), The Establishment of the Han Empire and Imperial China, Greenwood Press, London.

Hirth, F. (1917), The Story of Chang Ch'ien, China's Pioneer in West Asia, Journal of the American Oriental Society, Vol. 37, pp. 89–152.

Joyner, H. (1978), The Recruitment, Organisation, and Control of Former Han Military Leadership, Unpublished Master's Thesis, Department of Far Eastern Languages and Civilisations, University of Chicago.

Kaufman, C. (1987), The evaluation of marketing in a society: The Han dynasty of ancient China. Journal of Macromarketing 7:52–64

Kenderdine, S. (1995), Bai Jiao 1 – the Excavation of a Song Dynasty Shipwreck in the Dinghai Area, Fujian Province, The International Journal of Nautical Archaeology, 24, 4, pp. 247–266.

Koo, T. (1920), The Constitutional Development of the Western Han Dynasty, Journal of the American Oriental Society, Vol. 40, pp. 170–193.

Kuhn, D. (2000), Liao Architecture: Qidan Innovations and Han-Chinese Traditions? T'and Pao, Second Series, Vol. 86, Fasc. 4/5, pp. 325–362.

Lee, H. (2004), The Emperors Lady Ghost-writers in Song-Dynasty China, Artibus Asiae, Vol. 64, No. 1, pp. 61–101.

Lehman, J. (2006), Intellectual Property Rights and Chinese Tradition Section: Philosophical Foundations, Journal of Business Ethics, 69:1–9

Levathes, L. (1994), When China Ruled the Seas: The Treasure Fleet of the Dragon Throne, 1405–33, Open Road Distribution, New York.

Lewis, M. (2009), China Between Empires: The Northern and Southern Dynasties, The Belknap Press of Harvard University Press, Cambridge, Massachusetts.

Li, R. (2009), Private Property Rights, Legal Enforcement and Economic Prosperity: The Fall of Early Civilised China and the Rise of the United Kingdom in the 18th–19th century Asian Science Journal, Vol. 5, No.10.

Lin, J. (1992), Chinese Double-entry Book-keeping Before the Nineteenth Century, The Accounting Historians Journal, The Academy of Accounting Historians, Vol. 19, No. 2, pp. 103–122.

Lin, Y. J. (1995), The Needham Puzzle: Why the Industrial Revolution Did 'Not Originate in China', Economic Development and Cultural Change, Vol. 43, No. 2, pp. 269–292.

Little, S. (1996), Economic Change in Seventeenth-Century China and Innovations at the Jingdezhen Kilns, Ars Orientalis, Vol. 26, pp. 47–54.

Loewe, M. (1986), The Former Han Dynasty, in Twitchett, D., and Loewe, M. (Eds), The Cambridge History of China – Volume 1, The Ch'in and Han Empires, 221BC–220AD, Cambridge University Press, Cambridge.

MacCormack, G. (2004), The Transmission of Penal Law (Lü) from the Han to the T'ang: A Contribution to the Study of the Early History of Codification in China, Revue Internationale des Droits de l'Antiquité LI (2004): 47–83.

McLeod, A. (2015), Philosophy in Eastern Han Dynasty China (25 – 220 CE), Philosophy Compass, 10/6, pp. 355–368.

Pai, H. (1992), Culture contact and culture change: The Korean peninsula and its relations with the Han Dynasty commandery of Lelang, World Archaeology, 23:3, 306–319.

Pei, C. (1999), The Origins of Mediation in Traditional China, Dispute Resolution, 54, 2.

Rawson, J. (1999), The Eternal Palaces of the Western Han: A New View of the Universe, Artibus Asiae, Vol. 59, No. 1/2, pp. 5–58.

Rowe, D. (2005), The Origins of Value, Goetzmann, W., and Rouwenhorst, G. (Eds), Global Association of Risk Professionals, Issue 6, pp. 28.

Scogin, H. (1990), Between Heaven and Man: Contract and the State in Han Dynasty China, 63 S. Cal. L. Rev 1325.

Shaffer, L. (1986), China, Technology and Change, World History Bulletin 4, No. 1, pp 1–6.

Shang Zhijun et al. (1989), Essentials of the literature of traditional Chinese drug therapy, Beijing: Kexue jishu wenxian chubanshe.

Sheng, A. (1990), Textile Use, Technology and Change in Rural Textile Production in Song China (960AD to 1279AD), PhD Dissertation, University of Pennsylvania.

Standaert, N. (2008), The Interweaving of Rituals, Funerals in the Cultural Exchange between China and Europe, University of Washington Press, Seattle.

Tsou, C. (1998), Science and Scientists in China, Science, New Series, Vol. 280, No. 5363, pp. 528–529.

van Someren, T.C.R., & van Someren-Wang, S. (2013), Chinese culture, strategy, and innovation. Innovative China: Innovation Race between East and West (pp. 2756). Berlin: Springer

Wang, C., Yan, S., Hou, Z., Fu, W., Xiong, M., Han, S., Jin, L., and Li, H. (2012), Present Y Chromosomes reveal the ancestry of Emperor Cao Cao of 1800 years ago, Journal of Human Genetics, 57, pp. 216–218.

Wang, L., Ling, W., and Needham, J. (1955), Horner's Method in Chinese Mathematics: Its Origins in the Root Extraction Procedures of the Han Dynasty, T'oung Pao, Second Series, Vol. 43, Livr. 5, pp. 345–401.

Wong, E. (2011), Taoism: An Essential Guide, Shambhala Publications Inc, Boston, Massachusetts.

Xiong, V. (1996), Innovations and Taoism under Tang Xuanzong, T'oung Pao, Second Series, Vol. 82, Fasc. 4/5, pp. 258–316.

Yu-ch'uan, W. (1949), An Outline of The Central Government of the Former Han Dynasty, Harvard Journal of Asiatic Studies, Vol. 12, No. 1 / 2, pp. 134–187.

Zheng, C., and Knobloch, Y. (1983), A Discussion of the History of the Gu Zheng, Asian Music, Vol. 14, No.2, Chinese Music History, pp. 1–16.

9

The Mongol Empire: 1206 AD to 1368 AD

When Temuchin, later to be recognised as Genghis Khan, was born in 1167, the Mongols were part of a small group of nomads amongst many. Yet, at the peak of the power of the Mongolian Empire that he would establish, a population of one million Mongolians would control two-thirds of the old world, including three major civilisations[1]; and a mobile pastoral society would become highly militarised.[2] His father, Yesugei Baghutur, was the Khan of the small Kiyad tribe of the Mongols.[3] At the time of the birth of Genghis Khan, his father was returning from a revenge military campaign against another group of Mongols, the Tartars.[4] Genghis Khan and his first wife, Borte became betrothed in 1175 AD, when he was only eight years old. Borte was a few years older than he at the time.

[1] Biran, M. (2013), The Mongol Empire in World History: The State of the Field, History Compass 11/11, pp. 1021–1033.

[2] Di Cosmo, N. (1999), State Formation and Periodization in Inner Asian History, Journal of World History, Vol. 10, No. 1, pp. 1–40.

[3] Lattimore, O. (1963), Chingis Khan and the Mongol Conquests, Scientific American, Vol. 209, No. 2, pp. 54–71.

[4] Ibid.

S. Ramesh, *The Rise of Empires*, https://doi.org/10.1007/978-3-030-01608-1_9

The Mongols were aware of Daoism, Confucianism, Buddhism, and Nestorian Christianity, but were not attached to any of these, preferring to base their beliefs on the souls of humans and animals.[5] When Genghis Khan died in 1227 AD, his successors continued to expand and refine the organisational and political structure of his empire.[6] This realpolitik was based on an empire built for trade and exchange without regard to culture, tribal affiliations, language, or religion.[7] It was the idea that the Mongol Empire was not just imperial in nature, but also an economy based on trade and the exchange of goods, which served as a driving force for the Mongols' ongoing and future conquests.[8] In other ways, the Mongol Empire served to make the world a smaller place, with its intercontinental trading links. It also acted to agglomerate peoples of the same or similar language, giving them a more coherent national identity. One example of this would be the Turkic speaking tribes in the empire.[9] By conquering Islamic peoples in the Middle East and elsewhere in the empire, the Mongols also helped in the spread of Islam to places where it had never existed before, such as China.[10] Another way in which the Mongols introduced cohesiveness into their empire was through the exchange of daughters.[11] In this way, the Mongols would marry Mongol women to the warriors of conquered tribes and, in turn, Mongol warriors would marry women from conquered tribes. When Genghis Khan had consolidated his power amongst the Mongolian tribes and conquered the Uighurs and the Kirghiz, only two other powerful states lay to his south and to west.[12] These states were China and Khorezm, respectively. In

[5] Humphrey, C., and Hurelbaatar, A. (2005), Regret as a Political Intervention: An Essay in the Historical Anthropology of the Early Mongols, Past & Present, No. 186, pp. 3–45.

[6] Buell PD. (2007), How Genghis Khan has changed the World? Center for East Asian Studies, Western Washington University. http://www.mongolianculture.com/How%20Genghis%20Khan%20Has.pdf.

[7] Kotkin, S. (2007), Mongol Commonwealth? Exchange and Governance across the Post-Mongol Space, Kritika Explorations in Russian and Eurasian History, Vol. 8, No. 3, pp. 487–531.

[8] Ibid.

[9] Ibid.

[10] Ibid.

[11] Ibid.

[12] Prawdin, M. (2017), The Mongol Empire: Its Rise and legacy, Routledge.

1209 AD, the western Xia Dynasty of China submitted to the power of the Mongols, becoming a vassal state.[13] However, it had to be subdued by military force in 1227 AD to ensure its loyalty to Genghis Khan and the Mongols.[14] In 1211, Genghis Khan took control of a Mongol army to conquer the Jin dynasty of China. So much of the areas in the north had been subdued by the Mongols that, by 1214 AD, the Jin Emperor Xuanzong moved the Jin capital to Kaifeng in the south of China.[15] Kaifeng itself fell in 1233 AD to Ogedei Khan, the third son of Genghis Khan.[16]Khorezm was defeated in 1221 AD after a four-month campaign in which the Mongols took Bukhara, Samarkand, and Urgench.[17] Between 1211 AD and 1225 AD, the climate of the steppes of Mongolia was wetter and warmer, resulting in an abundance of food, which resulted in an enhancement of the concentration of military resources.[18] More births increased the manpower available to the Mongols to conduct warfare. During Genghis Khan's early teen years, the climate in Mongolia was less welcoming and featured several years of drought.[19] This resulted in conflict between the nomadic tribes of the steppes over limited resources, providing a background for the emergence of Genghis Khan as the unifier of the Mongolian people.

The expansion of the Mongols from the Steepes of Mongolia to the edge of Europe, Middle East, Central Asia, China, and Siberia brought together the cultures of the east and the west under one 'umbrella' for the first time in world history. Despite their conquests, the Mongols retained their Mongolian identity, but benefited from an exchange of ideas with other cultures and civilisations.[20] An Uighur seal-keeper who had worked

[13] Hung, M. (2016), From the Mongols to the Ming Dynasty: How a Begging Monk Became Emperor of China, Zhu Yuan Zhang, Algora Publishing, New York.

[14] Ibid.

[15] Ibid.

[16] Ibid.

[17] Prawdin, M. (2017), The Mongol Empire: Its Rise and legacy, Routledge.

[18] Jackson, P. (2017), The Mongols and the Islamic World – From Conquest to Conversion, Yale University Press, New Haven.

[19] Ibid.

[20] Schurmann, H. (1956), Mongolian Tributary Practices of the Thirteenth Century, Harvard Journal of Asiatic Studies, Vol. 19, No. 3/4, pp. 304–389.

for the ruler of the Turkic Naiman tribe, worked at the court of Genghis Khan after the Mongols had conquered the Naiman.[21] It was this Uighur seal-keeper who introduced the 14-letter Uighur alphabet in to the fledgling Mongolian state.[22] From the Khitans, who had ruled northern China and most of Mongolia from the tenth to the twelfth century, the Mongols received seasoned administrators who opted to work for them.[23] Furthermore, the Khitans were knowledgeable about how a state should be run. They, therefore, provided the Mongols with knowledge about the protocols and institutions that the Mongolian Empire would need for stability and growth.[24] Some argue that the Mongol Empire lacked the formal institutions of political power.[25] However, the protocols would have been associated with how a conquered state should be administered and taxed through the institution of a court-appointed Governor.[26] Mongolian taxation was of two types.[27] The first was a tribute, an annual percentage payment of the product of the common Mongolian and the subject vassal to his Lord. The second type of tax was a levy, the same as a tribute, but applied whenever extra revenue was needed. The other institutions of governance would have included a codified legal system, the Yasa, a hierarchy of political power, Khan, Council, nobility, and common folk. So, instead of saying that the Mongols did not have the formal institutions of political power, perhaps it would be more appropriate to say that these institutions did exist, but were less formal because the Khan's authority was the state.

The Khitans and the Uighurs were closely related to the Mongols, although this is truer of the Khitans, a proto-Mongol race, rather than of

[21] Morgan, D. (1982), Who ran the Mongol Empire?', JRAS 124–36).

[22] Blunden, J. (2004), Mongolia – the Bradt Travel Guide, The Globe Pequot Press Inc., USA.

[23] Morgan, D. (1982), Who ran the Mongol Empire?', JRAS 124–36).

[24] Ibid.

[25] Skrynnikova, T. (2011), Mongolian Nomadic Society of the Empire period, IN: Alternatives to Social Evolution, Kradin, N., Korotayev, A., Bondarenko, D. (Eds), LAP Lambert Academic Publishing.

[26] Morgan, D. (1982), Who ran the Mongol Empire?', JRAS, pp. 124–36.

[27] Smith, J. (1970), Mongol and Nomadic Taxation, Harvard Journal of Asiatic Studies, Vol. 30, pp. 46–85.

the Uighurs.[28] However, although the Mongols tried, Japan, Burma, and Java were never conquered.[29] Furthermore, while Vietnam was invaded on three occasions by the Mongols, it was never conquered. However, the Mongols did stimulate an increase in Vietnamese nationalism.[30] The rise and the fall of the Mongol Empire also saved as the force for awakening Europe from its long and primitive seclusion.[31]

The Mongols and Khorezm

The Khorezm civilisation was based around the banks of the River Amudar'ya, in what is now present-day Turkmenistan. The first Neolithic sites appeared in the Khorezm region in about 6000 BC, and the area continued to support the hunter-gatherer existence of people until 7 BC.[32]After this year, fortified sites began to appear; and between 6 BC and 2nd BC the region was known as Khorezm and it was a vassal state of the Persian Achaemenid Empire.[33] Archaeological evidence suggests that ancient Khorezm had many fortified settlements, villages, and an elaborate irrigation system using canals to divert water from the oasis to agricultural land.[34] However, after 1 BC, Khorezm became an independent state, having freed itself from Persian rule.[35] When the armies of Alexander the Great passed through Central Asia, Khorezm was neither assaulted nor conquered because Alexander the Great was heading for India.[36] Following the death of Alexander the Great, nomadic tribes living on

[28] Morgan, D. (1982), Who ran the Mongol Empire?', JRAS, pp. 124–36.

[29] Buell PD. (2007), How Genghis Khan has changed the World? Center for East Asian Studies, Western Washington University. http://www.mongolianculture.com/How%20Genghis%20Khan%20Has.pdf.

[30] Ibid.

[31] Ibid.

[32] Yagodin, V., and Betts, A. (2006), Ancient Khorezm, UNESCO.

[33] Ibid.

[34] Bolelov, S. (2016), Boris Vasilevich and the Study of Irrigation in Ancient Khorezm, IN Ancient Irrigation Systems of the Aral Sea Area: The History, Origin, and Development of Irrigated Agriculture, Andrianou, B., and Mantellini, S. (2016), Oxbow Books, Oxford.

[35] Yagodin, V., and Betts, A. (2006), Ancient Khorezm, UNESCO.

[36] Ibid.

China's frontiers began moving into the Samarkand region.[37] Over time, out of this movement of peoples emerged the Kushan Empire, based in northern Afghanistan around 1 BC.[38] The power of the Kushan Empire declined with the rise of the Sasanians in Persia.[39] The region was then invaded by the Huns, the Turks, the Arabs who brought Islam to the region, and finally, by the Mongols. By the time of the arrivals of the Mongols, Khorezm had an established Islamic tradition, with the region passing from the Afrighids (9 AD to 11 AD), Khorezmshah (12 AD to 13 AD), and finally, the Golden Horde (13 AD to 14 AD).[40] Khorezm under the Khorezmshah was a prosperous and powerful Islamic kingdom. Trade routes—north and south, east and west—were revived, and Khorezm prospered. As a result, there was also scholarship—the word 'algorithm' can be traced to Khorezm.[41] Khorezm was an empire that stretched from Iraq to Azerbaijan to eastern Afghanistan and, by the end of the twelfth century, its capital city was at Gurganj.[42] However, the Mongols would decimate the population of the conquered settlements and cities of Khorezm, but only after shipping back to Mongolia those who would be of worth—such as artisans.

Genghis Khan had initially wanted to maintain trading and commercial contacts with the Khorezmshahs.[43] However, in 1218 AD, a Mongol envoy travelling with some merchants was put to death alongside the merchants.[44] This brought the wrath of Genghis Khan down on the Khorezmshahs and Khorezm. At the same time, Genghis Khan sent the Mongolian cavalry under Subodai and Jebe, his two generals, towards southern Russia and the Caucasus, where they could win battles.[45]

---

[37] Ibid.

[38] Ibid.

[39] Ibid.

[40] Yagodin, V., and Betts, A. (2006), Ancient Khorezm, UNESCO.

[41] Metcalfe, D. (2009), Out of Steppe, Arrow Books, London.

[42] Ibid.

[43] Brummell, P. (2005), Turkmenistan – The Bradt Travel Guide, The Globe Pequot Press Inc., USA.

[44] Ibid.

[45] Buell PD. (2007), How Genghis Khan has changed the World? Center for East Asian Studies, Western Washington University. http://www.mongolianculture.com/How%20Genghis%20Khan%20Has.pdf.

The Mongols and China

In China, the Mongols conquered the northern Jin dynasty by 1233 AD. Kublai Khan was appointed as the Viceroy of northern China in 1251 AD by his eldest brother, the 4th Great Khan of the Mongol empire. Kublai Khan was the grandson of Genghis Khan and the son of Tolui and his wife Sorkaktani, from the Turkish speaking Kerait tribe.[46]Upon the death of his older brother, Kublai became the 5th Great Khan of the Mongol Empire, in 1260 AD. Kublai proclaimed the beginning of the Yuan dynasty in 1271 AD. And he had built his capital Dadu, which was to become Beijing, in 1267 AD.[47] Dadu was built according to plans very similar to the ones used by preceding non-Chinese dynasties to build their own capital cities.[48] The Mongols sacked the capital of the Jin dynasty, Zhongdu, in 1234 AD. The Jin had based their capital on the capitals of the Liao dynasty (Nanjing) and the Northern Song (Bianjing in Henan province), whom they had defeated in 1125 AD and in 1126 AD, respectively.[49] However, it would take another eight years before the Song armies could be finally crushed. Kublai Khan was the first emperor of a united China (north and south) at the start of the Yuan dynasty. This was accomplished in 1279 AD, when Lin'an, the capital of the Song dynasty, was captured by the Yuan army, and the Southern Song dynasty was crushed.[50] To effectively rule and control a population of 60–70 million Han Chinese by much fewer Mongols, Kublai centralised imperial administration, with the establishment of a secretariat that would enact laws and disburse punishments.[51] Kublai also retained the services of Confucian scholars in administrative positions. But the Chinese were

[46] Smith, J. (2000), Dietary Decadence and Dynastic Decline in the Mongol Empire, Journal of Asian History, Vol. 34, No. 1.

[47] Lim, S. (2011), Asian Civilisations – Ancient to 1800 AD, Asiapac, Singapore.

[48] Steinhardt, N. (1983), The Plan of Khubilai Khan's Imperial City, Artibus Asiae, Vol. 44, No. 2/3, pp. 137–158.

[49] Ibid.

[50] Lim, S. (2011), Asian Civilisations – Ancient to 1800 AD, Asiapac, Singapore.

[51] Ibid.

distrusted, and thus, prevented from having any political capability.[52] The Mongols facilitated the emergence of China as a unified state, which succeeding dynasties would preserve and extend. However, despite the Mongol tendency towards more centralised control, the controls were not tight and there was no accountability.[53] In contrast, the Uighurs and the Khitans had been more hierarchical and politically centralised.[54] The lack of accountability in Mongol central control led to the Mongols being easily bribed and deceived, leading to endemic corruption in the Mongol rule of China.[55] The Mongols were susceptible to being corrupted because of their love of decadence and profit.[56] However, despite these failings, the Mongols' success was due to their ability to mobilise human and material resources at will and on a massive scale.[57]

Tughon Temur was the last and the longest reigning Mongol Emperor of the Yuan dynasty in China.[58]However, due to revolts against his rule in China and natural disasters and plague, the rule of the Yuan dynasty in China became weakened, and it ended in 1368, when Tughon Temur was forced to flee China and return to Mongolia.[59] Like the other Mongolian elite of the Yuan dynasty, its last Emperor sought refuge in the central Mongolian city of Kharakhorum.[60] Although Mongol rule had ended, the institution of the 'Emperor' would continue to enshrine the principle

---

[52] Ibid.

[53] Dardess, J. (1978), Ming T'Ai-Tsu on the Yuan: An Autocrat's assessment of the Mongol Dynasty, Bulletin of Sung and Yuan Studies, No. 14, pp. 6–11.

[54] Rogers, J. D. (2015), Empire dynamics and Inner Asia. IN: J. Bemmann & M. Schmauder (Eds.), Complexity of interaction along the Eurasian steppe zone in the first millennium (pp. 73–88). Bonn: Bonn University Press.

[55] Ibid.

[56] Ibid.

[57] Jackson, P. (2000), The State of research: The Mongol Empire, 1986–1999, Journal of Medieval History, Vol. 26, No. 2, pp. 189–210.

[58] Morgan, D. (2009), The Decline and Fall of the Mongol Empire, JRAS, Series 3, 19, 4, pp. 427–437.

[59] Ibid.

[60] Honeychurch, W., and Amartuvshin, C. (2006), States on horseback: the rise of inner Asian confederations and empires. IN Archaeology *of Asia*, Stark, M. (Ed), PP. 255–278, London: Blackwell.

of Cakravartin in the subsequent Ming and Qing dynasties.[61] An official history of the Yuan dynasty, the Yuanshi, was written after the Mongols had been driven from China.[62] Its main themes: the identity of the Mongols related to military rhetoric and the notion of kingship legitimated by heaven.[63] The Qing dynasty also drew on the religious practices of the Mongols for their own court practices.[64]

China and Persia were the first of the four Khanates to become detached from Mongol control. It has been noted that both China and Persia were much more urbanised and sophisticated than were other contemporary societies.[65] In contrast to these urbanised and sophisticated societies, the Mongols were a nomadic people. In this case, it is evident that the Mongols found it difficult to manage and control urbanised societies.[66] However, through their military prowess, the Mongols were the architects of the world's first era of globalisation. In the context of China, Kublai Khan was probably the most international and outward looking of all the Yuan Emperors. During his reign, merchants from Persia would sell Persian goods in the markets of China and Persian engineers would design irrigation systems for Chinese cities and fields.[67] The Silk Route from China, through Central Asia, to Europe would flourish for the last time in the mid-thirteenth century. The Mongol Empire was the last imperial power to be able to give its protection to this ancient trade route. Furthermore, during the time of the Mongol Empire, the Silk Route had to pass through the Mongol capital, Karakorum, deep inside Mongolia.[68]

---

[61] Crossley, P. (1992), The Rulership's of China, American Historical Review, Vol. 97, No. 5, pp. 1468–1483.

[62] Dardess, J. (1978), Ming T'Ai-Tsu on the Yuan: An Autocrat's assessment of the Mongol Dynasty, Bulletin of Sung and Yuan Studies, No. 14, pp. 6–11.

[63] Francesca F. (2014), The Borders of Rebellion: The Yuan Dynasty and the Rhetoric of Empire, IN: Fiaschetti, F., and Schneider, J. (Eds), Political Strategies of Identity Building in non-Han empires in China (Asiatische Forschungen), Harrassowitz, Wiesbaden.

[64] Waley-Cohen, J. (2004), The New Qing History, Radical History Review, Issue 88, pp. 193–206.

[65] Marshall, R. (1993), Storm from the East, From Genghis Khan to Khubilai Khan, University of California Press, Berkeley, Los Angeles.

[66] Ibid.

[67] Ibid.

[68] Christian, D. (2000), Silk Roads or Steppe Roads? The Silk Roads in World History, Journal of World History, Vol. 11, No. 1, pp. 1–26.

The Persians also gained knowledge from China, especially with regards to ceramics and painting.[69] But whereas the Persian and Islamic connection with China began to lessen after the reign of Kublai Khan, Europe began to learn more from China, leading it into a period of conquest and innovation that lasts to this day. The North American landmass was encountered by Christopher Columbus in 1492 because he had set forth to discover a sea route to Kublai Khan's fabled Cathay.[70] Furthermore, the unilateral flow of knowledge and trade between China and Europe was such that it has been suggested that bubonic plague originated in China, but was 'transported' to Europe through Mongolian trade routes.[71]

The Mongols and Persia (Iran) and Iraq

Following the onslaught against the Khwarazmian Empire, the last ruler, Jalal-ad-Din, fled to northern India. Once Genghis Khan had returned to Mongolia, Jalal-ad-Din began the reconstruction of the Khwarazmian Empire.[72] However, Jalal-ad-Din could not return to the previous regions or cities he had controlled because these were now under Mongol control. In this case, Jalal-ad-Din seized Isfahan and Jibal in western Persia.[73] Jalal-ad-Din then went on to conquer Azerbaijan, while also militarily engaging the Christian kingdoms of Transcaucasia.[74] Because of his military engagements, Jalal-ad-Din conquered the whole of Persia by 1226 AD.[75] He then raided Georgia, which had become a Mongol vassal state, in 1226 AD, and again in 1228 AD.[76] This was then followed by a thrust

[69] Marshall, R. (1993), Storm from the East, From Genghis Khan to Khubilai Khan, University of California Press, Berkeley, Los Angeles.

[70] Ibid.

[71] Ibid.

[72] Sicker, M. (2000), The Islamic World in Ascending: From the Arab Conquests to the Siege of Vienna, Praeger Publishers, Westport.

[73] Ibid.

[74] Ibid.

[75] Ibid.

[76] Ibid.

into the territory of the Ayyubids of Syria, who as a result formed an alliance with the Seljukids of Damascus.[77]This coalition of Ayyubids and Seljukids defeated a military incursion into Anatolia by Jalal-ad-Din's armies.[78] In 1229 AD, the Great Khan Ogodei sent Mongolian troops, under the control of General Chormaghun, to quell the possibility of a resurgent Khwarazmian Empire.[79] Over time, Mongol forces took control of Azerbaijan, Greater Armenia, western and southern Persia, as well as Georgia in 1239.[80] In 1253 AD, the Great Khan Mongke sent his brother Hulegu to lead the Mongol armies in Persia and Iraq.[81] Hulegu conquered Alamut in Persia in 1256 AD, and Baghdad in 1258 AD, going on to conquer northern Syria but returning to Mongolia in 1260 AD to elect a new Great Khan.[82] This was to be his brother Kublai, who would also found the Yuan dynasty in China and become its first Emperor. The forces which Hulegu left in northern Syria were to be defeated by the Malmuks at the Battle of Ain Jalut, ending the Mongols' ambitions of gaining a foothold in either Egypt or Africa. In 1260 AD, following the election of Kublai as the Great Khan, the Mongol Empire was split into four and Hulegu took the Ilkhanate, which covered western Asia and Iraq.

In their conquests, the Mongolians used what was termed a 'Tsunami Strategy', in which a large Mongol force invaded a region and devastated it, but then drew back, retaining only a small part of the devastated region within their empire.[83] However, before the invasion, the Mongols would gather intelligence on the location of towns, military garrisons, the lay of the land, the culture, and the religion of the population from merchants

[77] Ibid.

[78] Ibid.

[79] Jackson, P. (2017), The Mongols and the Islamic World – From Conquest to Conversion, Yale University Press, New Haven.

[80] Ibid.

[81] Christie, N. (2014), Muslims and Crusaders: Christianity's Wars in the Middle East, 1095–1382, From The Islamic Sources, Routledge, New York.

[82] Ibid.

[83] May, T. (2016), Mongol Conquest Strategy in the Middle East, IN: The Mongols Middle East – Continuity and Transformation in Ilkhanaid Iran, De Nicola, B., and Melville, C. (Eds), Brill, Leiden.

and traders. The Mongols would then enter the region in a large force and attack the most populous, prosperous town before the force was divided to continue to ravage and devastate the entire region. The reason for the division of the initial invading force into many columns was to prevent the opposition from regrouping in large numbers.[84] The objective of the Mongols was to destroy the defensive and offensive military capabilities of the opposition, not to make the invaded region uninhabitable.[85] Thus, once this objective had been achieved, the Mongols would withdraw, without invading the whole region, and withdrawing to areas which had become stabilised.[86] The Mongols incorporated territories into their empire incrementally, without a sudden acquisition of territory.[87] However, the advance of the Mongols was rapid because once a stabilised region can be incorporated into their empire, they moved invaded unoccupied areas whose defences had been neutralised.[88] Furthermore, the Mongols used sophisticated signalling techniques on the battlefield. In conjunction with their policy of acquiring substantive intelligence about the area they were invading, the Mongols were never surprised.[89] The Mongols also invaded in large numbers, and their soldiers were highly mobile. So even if one column was defeated by an enemy, that enemy would have to face another Mongol column very soon after, without having the opportunity to rest its soldiers or replenish supplies.[90] The most important aspect of the stabilisation of conquered territory, and their incorporation into the empire was the transition from military to civilian rule.[91] Associated with this transition, the Mongols used a governing institution called a 'tamma'. This was a governing council in which the

[84] Ibid.

[85] Ibid.

[86] Ibid.

[87] Ibid.

[88] Ibid.

[89] May, T. (2016), Mongol Conquest Strategy in the Middle East, IN: The Mongols Middle East – Continuity and Transformation in Ilkhanaid Iran, De Nicola, B., and Melville, C. (Eds), Brill, Leiden.

[90] Ibid.

[91] Ibid.

nucleus was the Mongolian nobility, supported by administrators from the local population.[92] The 'tamma' also included the main military advance force and the scouts.[93] While the Mongols did not build fortified garrisons in conquered territories, they razed all such existing fortifications so they could not be used against them.[94] Furthermore, the Mongols used 'tamma' camps at the frontiers of their empire, in conquered territories, in order to defend and expand the borders of their empire.[95] Over time, the governance structure in conquered territories would change from a 'tamma' to a governor, secretary, and administrator, who would be responsible for collecting taxes and for maintaining local law and order.[96] The inclusion of a governing body within the main invading army allowed for flexibility in which the army was used. The 'tamma' would be used to stabilise and administer invaded regions that had come to normality, while the main body of the army could be diverted to areas in which revolts were occurring.[97]

The Mongols and Europe

The first Mongolian foray into Europe occurred in 1221, under their generals Subodei and Jebe. The Mongolian incursion into Europe, whose inhabitants knew little if anything of the Mongol hordes, occurred following Genghis Khan's invasion of Central Asia and the demise of the Khwarizm sultanate.[98] Once this had been accomplished, Genghis Khan gave his permission to Subodei and Jebe to take their armies to the north and into Europe, in what then was unknown territory for the Mongols. The first Christian kingdom the Mongols found was the small kingdom

[92] Ibid.

[93] Ibid.

[94] Ibid.

[95] Ibid.

[96] Ibid.

[97] Ibid.

[98] Weatherford, J. (2004), Genghis Khan and the Making of the Modern World, Crown Publishers, New York.

of Georgia, which was quickly subdued, becoming a Mongolian vassal state in the process.[99] Once a state attained vassal status under the Mongols, they were saved from any military attack by the Mongols if they continued to pay tribute. For example, when the Tibetans stopped paying tribute to the Mongols, an army was sent in 1240 AD, 1244 AD, and 1251 AD.[100]

Following the subjugation of Georgia, the Mongols then moved on into the flat lands of eastern Europe in order to reconnoitre and gather information about the size of the population, the number and the location of cities, as well as their political affiliations.[101] On the way, Subodei and Jebe encountered a nomadic tribe called the Kipchak, some of whom were persuaded by the Mongols to become their allies.[102] The Mongolians also made a pact with Venetian merchants in which the Venetians would send back information to the Mongols about the peoples of the lands in which they traded.[103] In return, the Mongolians gave the Venetians monopoly trading rights in their lands with the promise of disposing of any rivals.[104] By April 1223 AD, the Mongols had reached the River Dnieper, by which time the local population had become aware of the invaders. As such, word reached the ears of the Christian city states in the paths around the invading Mongol armies. These Christian city states included Smolensk, Galich, Chernigov, Kiev, Volhynia, Kursk, and Suzdal.[105] As well as a proportion of the Kipchaks, whose allegiance had not been won by the Mongols. While the Christian city states sent out armies to defeat the Mongols, the Christian armies were themselves outmanoeuvred and crushed by the Mongolian guruns. The Mongols left Europe in 1224 AD, 3 years after they arrived, on their way to Mongolia

[99] Ibid.

[100] Wylie, T. (1977), The First Mongol Conquest of Tibet Reinterpreted, Harvard Journal of Asiatic Studies, Vol. 37, No. 1, pp. 103–133.

[101] Ibid.

[102] Ibid.

[103] Gabriel, R. (2006), Genghis Khan's Greatest General Subotai the Valiant, University of Oklahoma, Norman.

[104] Ibid.

[105] Weatherford, J. (2004), Genghis Khan and the Making of the Modern World, Crown Publishers, New York.

via the banks of the Syr Darya River, where they would be met by Genghis Khan himself.[106] However, in 1237 AD, the Mongol armies returned to start the conquest of southern Russia, and then, entered Poland and Hungary in 1241 AD.[107] However, news reached the Mongol armies of the death of Ogodei, the Great Khan, soon afterwards, and they headed back to Mongolia in 1241 AD.[108] Hundreds and thousands of people were killed by the Mongols, in eastern Europe, Central Asia, and in other parts of the Mongol Empire. Europe would not get to witness such barbarity until the third and the fourth decade of the twentieth century, when Hitler would unleash a similar level of barbarity.[109]

It was due to the Mongols that the knowledge of printing, gunpowder, and the compass diffused from China to Europe.[110] Knowledge of these three Chinese inventions would catapult the European powers to world conquest and dominance over the following centuries. In 1300 AD, Marco Polo's travel logs, 'The Travels of Marco Polo', became available. The book describes Polo's travels in Asia as well as his experiences at the court of Kublai Khan, between 1271 AD and 1295 AD. He set out from Venice in 1271 AD on his travels in the east, at the age of eighteen with his father and uncle, and ended up in the court of the Great Khan, Kublai. While Polo, his father and uncle may have served as auditors for the Mongols, he also claims that he had served in an administrative and diplomatic role for the Mongols.[111] This claim may not be as farfetched because Polo would have been able to learn to speak, and maybe to write Mongolian, Chinese, and Persian.[112] The first language would have been

[106] Gabriel, R. (2006), Genghis Khan's Greatest General Subotai the Valiant, University of Oklahoma, Norman.

[107] Pacey, A. (1990), Technology in World Civilisation: A Thousand Year History, The MIT Press, Cambridge, Massachusetts.

[108] Grant De Pauw, L. (1998), Battle Cries and Lullabies, Women in War from Prehistory to the Present, University of Oklahoma Press, Norman.

[109] Breitman, R. (1990), Hitler and Genghis Khan, Journal of Contemporary History, Vol. 25, pp. 337–351.

[110] Weatherford, J. (2004), Genghis Khan and the Making of the Modern World, Crown Publishers, New York.

[111] Wolfe, A. (2014), Marco Polo: Factotum, Auditor, Language and Political Culture in the Mongol World Empire, Literature Compass, 11/7, pp. 409–422.

[112] Ibid.

the most important, as it would have allowed Polo to act as a functionary for the Emperor.[113] Polo was 42 when he returned home in 1295 AD.[114] During the years he spent in the court of Kublai, Polo would have been able to acquire knowledge about the technology to which the Mongols and their subjects would have access. In this case, Polo may have brought back with him to Europe specific knowledge of gunpowder, printing, the compass, and the ships used by the Chinese. He had also witnessed the use of paper money in Yuan China in 1298 AD.[115] The Europeans then could develop these Chinese inventions and design and build ships that were more efficient at taking advantage of the winds to take them far from their European homelands, with the assistance of the compass.

The Mongol Empire and Innovation

The Mongolians were not technical, agricultural, or cultural innovators. They did not establish any religion, and neither were they craftsmen.[116] In this case, the Mongols were not bakers, makers of pottery, metal workers, artists, or weavers.[117] And the Mongols were neither scholars nor prolific writers, although they did use the services of scholars in the domains they had conquered, when it was necessary. One of these occasions was the writing of the 'Secret History of the Mongols' in 1240 AD.[118] Although the original version in the Mongolian language disappeared many centuries ago, copies of the book in the Chinese script have survived into recent times, and have been used by contemporary scholars to study the Mongol Empire. However, while the Mongols may not have been as innovative as either past or contemporary civilisations were, the

[113] Ibid.

[114] Ibid.

[115] Gunaratne, S. (2001), Paper, Printing and the Printing Press, Gazette Vol. 63 (6), pp. 459–479.

[116] Weatherford, J. (2004), Genghis Khan and the Making of the Modern World, Crown Publishers, New York.

[117] Ibid.

[118] Lattimore, O. (1963), Chingis Khan and the Mongol Conquests, Scientific American, Vol. 209, No. 2, pp. 54–71.

founder of the Mongol Empire, Genghis Khan, was a tenacious military genius with a formidable ability to organise and structure his forces in such a way that the Mongols were able to use their cavalry, foot soldiers, and archers to defeat more innovative civilisations. Not only Genghis Khan, but his generals, Subedei and Jebe, were masters of the art of battlefield strategy, tactics, and operations.[119] All three were versed in the art of stealth and surprise, of the need to meet and defeat the enemy quickly by passing the islands of resistance (walled cities), and of the need to live off the land to ensure quick mobility.[120] The Mongol warrior was typically dressed in armour that constituted numerous layers of bull hide, which afforded the warrior secure and impenetrable protection.[121] Furthermore, Mongol arrowheads were typically longer than those of the Europeans, for example.[122] Moreover, Mongol arrowheads were made of iron, bone, or horn and, as the slit ends were bound tightly, Mongol arrows could not fit European bows.[123] Therefore, Mongol arrows could have greater range and were more deadly than the arrows of their enemies. Furthermore, Mongol arrows could not be reused by their enemies in battle. However, Mongol warfare was given a competitive advantage by its tenacity and its sheer size.[124]

The Mongols were also disciplined and, to some extent, this was the second occasion on which a written document was produced. This document, the Yasa, represented 22 laws which Genghis Khan laid down in 1206 AD.[125] The Yasa were laws that bound the Mongols, and these laws were based on a combination of the ancestral customs and laws of the

---

[119] Bellamy, C. (1983), Heirs of Genghis Khan: The Influence of Tartar Mongols on the Imperial Russian and Soviet Armies, The RUSI Journal, 128, 1, pp. 52–60.

[120] Ibid.

[121] Sweeney, J. (1982), Thomas of Spalato and the Mongols: A Thirteenth-Century Dalmatian View of Mongol Customs, Florilegium, Vol. 4, pp. 156–183.

[122] Ibid.

[123] Sweeney, J. (1982), Thomas of Spalato and the Mongols: A Thirteenth-Century Dalmatian View of Mongol Customs, Florilegium, Vol. 4, pp. 156–183.

[124] Smith, J. (1975), Mongol Manpower and Persian Population, Journal of the Economic and Social History of the Orient, Vol. 18, No. 3, pp. 271–299.

[125] Morgan, D. (1986), 'The "Great Yasa of Chingiz Khan" and Mongol Law in die Ilkhanate', *Bulletin of the School of Oriental and African Studies*, xlix, pp. 169–70.

Mongols, as well as new laws devised by Genghis Khan himself.[126] The laws related to how subject peoples and property should be divided up amongst the Mongolian nobility and what crimes resulted in death or some other form of punishment.[127] In addition to the Yasa, there was also another document, the Bilik, which represented Genghis Khan's suggestions and comments on various issues.[128] Genghis Khan was illiterate, so he probably expressed himself to scribes, who would have written his words using either the Mongol or Uighur script in the Yasa and the Bilik.[129] The Yasa was made widely available throughout the empire, and even the library at Baghdad had a copy.[130] Furthermore, the Yasa may have been modified as the Mongol Empire expanded into China in 1211 AD and Turkestan in 1219 AD, with the help of Persian, Chinese, and Uighur scholars.[131] However, the Yasa only survived in fragments, and any interpretation of it is based on second accounts by other scholars.[132]

Thus, it was the military innovation and the discipline of the Mongols which enabled their prowess to defeat more innovative and established civilisations (Khorezm, China, Persia, Iran), as well as less advanced civilisations (Europe). However, the Mongols also had the ability to acquire knowledge about the peoples and the practices in specific regions before they set forth on a military campaign. For example, in their conquest of Georgia, they had known that the local practice was to carry Christian crosses—by the local military and the people. So, the invading Mongol army carried a cross in front of it as it marched through Georgia.[133] The local population was beguiled into thinking that the Mongols were friendly and Christian and so, on seeing the approaching

[126] Ibid.

[127] Ibid.

[128] Bartold, V. (1928), Turkestan down to the Mongol Invasion, London, 1928.

[129] Vernadsky, G. (1938), The Scope and Contents of Chingis Khan's Yasa, Harvard Journal of Asiatic Studies, Vol. 3, No. 3/4, pp. 337–360.

[130] Ibid.

[131] Ibid.

[132] Ibid.

[133] Jackson, P. (2005), The Mongols and the Faith of the Conquered, IN: Mongols, Turks and Others – Eurasian Nomads and the Sedentary World, Amitai, R., and Biran, M. (Eds), Koninklijke Brill NV, Leiden, The Netherlands.

Mongol army, the city gates would be flung open and people would approach the Mongols with crosses in their hands, only to be massacred. And, as the frontiers of the Mongolian Empire continued to grow and expand, the Mongols built transparent and non-transparent bridges to connect the civilisations and the cultures they had conquered.[134] This created a cauldron in which a diverse range of innovations could mix and generate further innovations. For example, the cannon emerged as result of the combination of innovations associated with Chinese gunpowder, the Muslim flamethrower, and European bell-making technology by engineers from China, Persia, and Europe.[135] The cannon was a cascading technology from which emerged the pistol, the gun, and ultimately, missiles.[136] However, the Mongols should not be viewed as being just the passive transmitters of religious, cultural, and technical knowledge. They were able to adapt such knowledge in order to suit their own tastes, needs, and circumstances.[137] Following the withdrawal of the Mongol armies following the conquest of Poland and Hungary in 1241 AD, the Europeans were intrigued by the Mongols, and in the lands from which they had come. As a result, a Papal ambassador was sent to Mongolia in 1245 AD, to be followed by William of Rubruck, who returned to Europe in 1257 AD.[138]Not soon after his return in 1258 AD, the first reports of the use of gunpowder in fireworks in the city of Cologne were to be written in Europe.[139] Two books about gunpowder weapons appeared in the eastern Mediterranean in 1280 AD.[140]

While the Mongols did not use gunpowder weapons in Hungary in 1241 AD, it appears that they acquired it not long after.[141] However,

---

[134] Weatherford, J. (2004), Genghis Khan and the Making of the Modern World, Crown Publishers, New York.

[135] Ibid.

[136] Ibid.

[137] Allsen, T. (1997), Commodity and Exchange in the Mongol Empire: a cultural history of Islamic Textiles, Cambridge University Press, Cambridge.

[138] Pacey, A. (1990), Technology in World Civilisation: A Thousand Year History, The MIT Press, Cambridge, Massachusetts.

[139] Ibid.

[140] Ibid.

[141] Ibid.

another source suggests that the Mongols did use gunpowder for military purposes at the Battle of Sajo in Hungary in 1241 AD.[142] The Mongols again made use of gunpowder for military purposes against Japan in 1274 AD and 1281 AD.[143] Since gunpowder and associated technology were invented in China, the Mongols acquired it from the Chinese. Gunpowder with less explosive power was used to propel rockets in 1150 AD.[144] However, by 1250 AD, the Chinese had developed a gunpowder formula with more explosive power. This is evidenced by the fact that, in that year, Mongolian armies invaded Iran and these armies took with them Chinese engineers who operated catapults which could fling explosive devices against any cities and armies in their path.[145] Nevertheless, there is some evidence to suggest that the development of gunpowder with such explosive force as to propel projectiles at an enemy had already led to such use at the siege of Kaifeng in 1126 AD.[146] It is also evident that, by 1249 AD, Islamic powers in the Middle East also had an awareness of gunpowder weapons, because such weapons were used to repulse Christian crusaders from Palestine in 1291 AD, following the siege of Acre.[147] Furthermore, the invading Mongol armies were also stopped by the Islamic use of gunpowder weapons when they entered Syria in 1258 AD.[148] However, the presence of Iranian engineers in the Mongol army, in the last stages of its campaign against the Song dynasty, between 1250 AD and 1276 AD,[149] led to the development and use of more sophisticated catapults to throw explosive devices more accurately and with greater impact. Furthermore, by 1280 AD, both Chinese and Islamic armies were using a crude form of gun called a 'fire-lance'.[150]By 1288 AD, Chinese soldiers

[142] Contamine, P. (1984), War in the Middle Ages. Basil Blackwell, Oxford.

[143] Ibid.

[144] Pacey, A. (1990), Technology in World Civilisation: A Thousand Year History, The MIT Press, Cambridge, Massachusetts.

[145] Ibid.

[146] Ibid.

[147] Ibid.

[148] Ibid.

[149] Ibid.

[150] Ibid.

in the Mongol army were using gunpowder weapons that had evolved further from the 'fire-lance'. This is evidenced by the fact that a bronze spherical barrel was found at the site of a battle in which the Mongolians fought, in 1288 AD.[151] Chinese gunpowder barrels evolved further by having thicker walls, to better contain the explosive power leading to greater projection of the projectile inside the barrel, and the explosion was caused by ignition at the base of the barrel.[152]

Although Genghis Khan did not change the Mongols' traditional nomadic principles of warfare, he did structure the Mongolian cavalry, as well as instil rigorous discipline into its ranks. He did this by winning the loyalty of his commanders as well as that of the common soldier. Officers had to treat soldiers fairly, booty was equally divided, and all soldiers were given the same food regardless of their rank.[153] Furthermore, Genghis Khan also used unconventional battle tactics. One way he did this was to divide the Mongolian cavalry into a 'guran', which was ten horsemen deep and a hundred horsemen wide.[154] Each 'guran' would train against the other, with regards to flanking manoeuvres and sham fights.[155] Furthermore, whenever the Mongols went on military campaigns, they would take very many horses, such that there were 20–30 horses per cavalryman.[156] No stores were taken for the horses by the Mongols who fed them on the grass of the plains where the Mongol army stopped. This gave the Mongol armies great flexibility to march over long distances whenever it was required to do so. Barrels also came to be fitted inside a frame, in order that the shot could be better stabilised. European drawings of gunpowder barrel guns in 1326 and 1327 show designs strikingly like the earlier Chinese ones.

In association with the political restructuring of their empire, the Mongols also created a common imperial culture which quickly spread

---

[151] Ibid.

[152] Ibid.

[153] Ibid.

[154] Prawdin, M. (2017), The Mongol Empire: Its Rise and legacy, Routledge.

[155] Ibid.

[156] Lattimore, O. (1963), Chingis Khan and the Mongol Conquests, Scientific American, Vol. 209, No. 2, pp. 54–71.

beyond the borders of the empire (Buell 2007). For example, the elements of Mongol culture included their dress style, their food, and the musical instruments they played. Typical Mongolian dishes included the Shulen (Mongol banquet soup), the Shaqimur (rape turnip soup), Eggplant manta, and Gullach. These dishes had health-promoting potential and were eaten in addition to animal body parts and herbs that were also thought to promote health.[157] A dietary manual, the 'Yinshan Zhengyao', was presented to the court of the Yuan Emperor in 1330 AD.[158] As the best part of their diet was liquid in form, the Mongols preferred the use of porcelain, because the glazed pottery would prevent the absorption of the liquid.[159] The Mongols favoured medicinal cures from the Middle East, and the composition of these were gathered in a manual called the 'Huihui Yaofang'.[160] The Mongolians also created innovative common protocols, such as the granting 'diplomatic immunity' rather than keeping hostages. Furthermore, in China, the Yuan dynasty and its government also undertook large-scale projects such as changing the course of the Yellow River so it flowed into the sea south of the Shandong peninsula.[161] This was a technological feat in itself.

End of Empire

The death of Mongke, the fourth Great Khan of Mongolia, in 1259, may have been the event that set in motion all the events which led to the effective dissolution of the Mongolian Empire.[162] In the choice of a Great Khan to succeed Mongke, Mongolian traditional customary practices

[157] Prawdin, M. (2017), The Mongol Empire: Its Rise and legacy, Routledge.

[158] Ibid.

[159] Prawdin, M. (2017), The Mongol Empire: Its Rise and legacy, Routledge.

[160] Ibid.

[161] Morgan, D. (2009), The Decline and Fall of the Mongol Empire, JRAS, Series 3, 19, 4, pp. 427–437.

[162] Jackson, P. (1978), The Dissolution of the Mongol Empire, Central Asiatic Journal, Vol. 22, No. 3/4, pp. 186–244.

came into conflict with the common-sense, seniority-based rules of succession.[163] In this case, the customary practice was that the third son of the chief wife of the ruling Great Khan should inherit his seat, or 'ordo'.[164] However, practicality emphasised the need for seniority to play a more significant role in the choice of a successor to the Great Khan. This conflict and practicality pitted Kublai Khan against his younger brother Ariq Boke, the sons of Tolui, Genghis Khan's fourth son. In 1260 AD, Kublai Khan was chosen by his people in a special meeting held in Kharakhorum, becoming the 5th Great Khan of the Mongolian Empire. The selection was contested by a rival, Ariq Boke, with whom Kublai fought many battles, but with Boke dying of a non-battle related illness in 1266 AD.[165] This followed five years of internecine fighting.[166] At the same time as the conflict between Kublai and his younger brother Ariq Boke, there was also a conflict between their brother Hulegu and their cousin, Berke the son of Jochi, Genghis Khan's eldest son and the brother of Tolui Khan. Hulegu was the Khan of the Ilkhanate, which initially encompassed Iran, Azerbaijan, and parts of contemporary Turkey, but later expanded to occupy neighbouring areas. Berke was the son of Jochi, who was the oldest son of Genghis Khan. Berke became Khan of the Golden Horde, an area encompassing parts of modern-day Russia and Central Asia, in 1255 AD, following the death of his brother Batu.[167] Berke had become the first Mongol to convert to Islam in the 1250s. Although his grandfather Genghis Khan had forbidden the Muslim practice of killing animals to be eaten according to Sharia law.[168] When his cousin, Hulegu, Khan of the Ilkhanate, executed the last Abbasid Caliph of Baghdad in 1258, Berke

[163] Ibid.

[164] Jackson, P. (1978), The Dissolution of the Mongol Empire, Central Asiatic Journal, Vol. 22 No. 3 /4, pp. 186–244.

[165] Rossabi, M. (2009), Khubilal Khan: His Life and Times, University of California Press, Berkeley.

[166] Jackson, P. (1978), The Dissolution of the Mongol Empire, Central Asiatic Journal, Vol. 22 No. 3/4, pp. 186–244.

[167] Mikaberidze, A. (2011), Conflict and Conquest in the Islamic World: A Historical Encyclopedia, ABC CLIO, Oxford, England.

[168] Jackson, P. (2005), The Mongols and the Faith of the Conquered, IN: Mongols, Turks and Others – Eurasian Nomads and the Sedentary World, Amitai, R., and Biran, M. (Eds), Koninklijke Brill NV, Leiden, The Netherlands.

was enraged.[169] However, it was not this, but the threat of the Ilkhanate expanding and taking further contested territory, the Caucasuses, Azerbaijan, Syria, and Egypt, which led to a military conflict between the Golden Horde and the Ilkhanate in 1262 AD.[170] This saved Europe from another Mongolian invasion, which Berke had been planning.[171] In contrast to the conversion of Berke, Hulegu seemed more drawn to Christianity. This is evidenced by a policy of co-operation between the Ilkhanate and the Pope in Rome emerging in 1264 AD.[172] However, Hulegu seems to have been drawn to the Christians from a strategic point of view rather than a faith-based one.

Since Kublai Khan also ruled northern China at the same time, he moved his capital from Kharakhorum in the Mongolian heartlands to Peking in China. This detracted from their traditional practice of ruling over their lands through the mobility of the 'gher', the circular, domed tent the Mongolians called home. The lack of a fixed capital, which other civilisations did have, had given the Mongols the ability to dispatch warriors and to move with speed whenever it was needed.[173] The year 1260 AD was a watershed year for the Empire founded by Genghis Khan and his descendants. Firstly, in 1260 AD, the Mongols were defeated in the Battle of Ain Jalut in Galilee by the Malmuks of Egypt.[174] This stopped the Mongols from conquering either Egypt or Africa.[175] Furthermore, after 1260 AD, the Mongol Empire split into four Khanates. Some attribute this fragmentation of the Mongol Empire directly to the fact that Kublai shifted his capital to Peking.[176] The four Khanates included

[169] Mikaberidze, A. (2011), Conflict and Conquest in the Islamic World: A Historical Encyclopedia, ABC CLIO, Oxford, England.

[170] Ibid.

[171] Morgan, D. (1989), The Mongols and the eastern Mediterranean, Mediterranean Historical Review, 4:1, pp. 198–211.

[172] Richard, J. (1969), The Mongols and the Franks, Journal of Asian History, Vol. 3, No. 1, pp. 45–57.

[173] Lattimore, O. (1963), Chingis Khan and the Mongol Conquests, Scientific American, Vol. 209, No. 2, pp. 54–71.

[174] Saunders, J. (1971), The History of the Mongol Conquests, University of Pennsylvania Press, Philadelphia.

[175] Ibid.

[176] Lattimore, O. (1963), Chingis Khan and the Mongol Conquests, Scientific American, Vol. 209, No. 2, pp. 54–71.

Mongol China under the Yuan dynasty (1260 AD–1368 AD), the Golden Horde which dominated western Siberia and Russia, the Khanate of Chaghadai which included Turkistan and the surrounding steepes, and lastly, the Ilkhanate, which incorporated Iran and Iraq.[177] The fragmentation of the Mongol Empire into the four Khanates led to less cohesion and loyalty between the members of the Mongol ruling family. As a result, the Khanates were less willing to support each other as each ruler took to a specific religion. The turnover of Mongol rulers in the Khanates was high, and compared to the Manchu dynasty in China, the Khanates did not survive for as long. In this case, the Manchu dynasty in China lasted for approximately two-and-a-half centuries—1644 to 1908—the four Khanates, in comparison, lasted for an average of 107 years.[178] Furthermore, whereas the Manchu dynasty had fewer Emperors with longer average reigns, the Khanates had more rulers who reigned for shorter periods of time. For example, the Yuan dynasty ruled China for 110 years with ten Emperors, the Golden Horde lasted for 132 years with twelve Khans, and Mongol rule in the Middle East lasted for 80 years with nine Mongol rulers.[179] These statistics would lend to the view that the Mongol rulers of the Khanates suffered high mortality rates due to a poor diet as well as a high level of alcoholism.[180] The Mongol diet mainly consisted of meat (lamb, goat, horsemeat) and their favourite drink was mare's milk, which, when fermented, would turn into alcohol.[181] In contrast, compared to prosperous Mongolians, poor Mongolians had a healthy diet, as Temuchin (Genghis Khan) discovered in the early years of his life, when he lived with his widowed mother and all the family's livestock had been stolen.[182] In this case, Temuchin's mother taught him how to survive by eating wild plants and hunting animals and fish as best he could. This ensured that they had a balanced diet, and less time and opportunity to engage in drinking fermented mare's milk.

---

[177] Prawdin, M. (2017), The Mongol Empire: Its Rise and legacy, Routledge.

[178] Smith, J. (2000), Dietary Decadence and Dynastic Decline in the Mongol Empire, Journal of Asian History, Vol. 34, No. 1.

[179] Ibid.

[180] Ibid.

[181] Ibid.

[182] Ibid.

The social, cultural, religious, racial, and political make-up of inner Asia was changed by the Mongol conquests in the east (China), south-west (Iraq and Iran), south (northern India), and the west (Russia and eastern Europe).[183] The Mongol conquests of the twelfth and thirteenth centuries also downgraded the importance and power of former trading centres such as Samarkand.[184] Furthermore, the end of the Mongol Empire gave way to the rise of the Ming dynasty in China, the Empire of Akbar in India, the Safavid Empire in Iran, and—even more game changing—the rise of the European powers and a new era of their imperial power and colonisation.[185] The most important and most consequential of the European powers to world history would be the British Empire. Without Genghis Khan and the Mongolian Empire, world history may not have reached the level of innovative uniqueness it has achieved in the twenty-first century.

References

Allsen, T. (1997), Commodity and Exchange in the Mongol Empire: a cultural history of Islamic Textiles, Cambridge University Press, Cambridge.

Bartold, V. (1928), Turkestan down to the Mongol Invasion, London, 1928.

Bellamy, C. (1983), Heirs of Genghis Khan: The Influence of Tartar Mongols on the Imperial Russian and Soviet Armies, The RUSI Journal, 128, 1, pp. 52–60.

Biran, M. (2013), The Mongol Empire in World History: The State of the Field, History Compass 11/11, pp. 1021–1033.

Breitman, R. (1990), Hitler and Genghis Khan, Journal of Contemporary History, Vol. 25, pp. 337–351.

Brummell, P. (2005), Turkmenistan – The Bradt Travel Guide, The Globe Pequot Press Inc, USA.

Blunden, J. (2004), Mongolia – the Bradt Travel Guide, The Globe Pequot Press Inc, USA.

---

[183] Franks, A. (1990), The Thirteenth-Century World System: A Review Essay, Journal of World History, Vol. 1, No. 2, pp. 249–256.

[184] Ibid.

[185] Ibid.

Bolelov, S. (2016), Boris Vasilevich and the Study of Irrigation in Ancient Khorezm, IN Ancient Irrigation Systems of the Aral Sea Area: The History, Origin, and Development of Irrigated Agriculture, Andrianou, B., and Mantellini, S. (2016), Oxbow Books, Oxford.

Buell PD. (2007), How Genghis Khan has changed the World? Center for East Asian Studies, Western Washington University. http://www.mongolianculture.com/How%20Genghis%20Khan%20Has.pdf

Christian, D. (2000), Silk Roads or Steppe Roads? The Silk Roads in World History, Journal of World History, Vol. 11, No. 1, pp. 1–26.

Christie, N. (2014), Muslims and Crusaders: Christianity's Wars in the Middle East, 1095–1382, From The Islamic Sources, Routledge, New York.

Contamine, P. (1984), War in the Middle Ages. Basil Blackwell, Oxford.

Crossley, P. (1992), The Rulership's of China, American Historical Review, Vol. 97, No. 5, pp. 1468–1483.

Dardess, J. (1978), Ming T'Ai-Tsu on the Yuan: An Autocrat's assessment of the Mongol Dynasty, Bulletin of Sung and Yuan Studies, No. 14, pp. 6–11.

Di Cosmo, N. (1999), State Formation and Periodization in Inner Asian History, Journal of World History, Vol. 10, No. 1, pp. 1–40.

Francesca F. (2014), The Borders of Rebellion: The Yuan Dynasty and the Rhetoric of Empire, IN: Fiaschetti, F., and Schneider, J. (Eds), Political Strategies of Identity Building in non-Han empires in China (Asiatische Forschungen), Harrassowitz, Wiesbaden.

Franks, A. (1990), The Thirteenth-Century World System: A Review Essay, Journal of World History, Vol. 1, No. 2, pp. 249–256.

Gabriel, R. (2006), Genghis Khan's Greatest General Subotai the Valiant, University of Oklahoma, Norman.

Grant De Pauw, L. (1998), Battle Cries and Lullabies, Women in War from Prehistory to the Present,' University of Oklahoma Press, Norman.

Gunaratne, S. (2001), Paper, Printing and the Printing Press, Gazette Vol. 63 (6), pp. 459–479.

Honeychurch, W., and Amartuvshin, C. (2006), States on horseback: the rise of inner Asian confederations and empires. IN Archaeology *of Asia*, Stark, M. (Ed), PP. 255–278, London: Blackwell.

Humphrey, C., and Hurelbaatar, A. (2005), Regret as a Political Intervention: An Essay in the Historical Anthropology of the Early Mongols, Past & Present, No. 186, pp. 3–45.

Hung, M. (2016), From the Mongols to the Ming Dynasty: How a Begging Monk Became Emperor of China, Zhu Yuan Zhang, Algora Publishing, New York.

Jackson, P. (1978), The Dissolution of the Mongol Empire, Central Asiatic Journal, Vol. 22, No. 3/4, pp. 186–244.

Jackson, P. (2000), The State of research: The Mongol Empire, 1986–1999, Journal of Medieval History, Vol. 26, No. 2, pp. 189–210.

Jackson, P. (2005), The Mongols and the Faith of the Conquered, IN: Mongols, Turks and Others – Eurasian Nomads and the Sedentary World, Amitai, R., and Biran, M. (Eds), Koninklijke Brill NV, Leiden, The Netherlands.

Jackson, P. (2017), The Mongols and the Islamic World – From Conquest to Conversion, Yale University Press, New Haven.

Kotkin, S. (2007), Mongol Commonwealth? Exchange and Governance across the Post-Mongol Space, Kritika Explorations in Russian and Eurasian History, Vol. 8, No. 3, pp. 487–531.

Lattimore, O. (1963), Chingis Khan and the Mongol Conquests, Scientific American, Vol. 209, No. 2, pp. 54–71.

Lim, S. (2011), Asian Civilisations – Ancient to 1800AD, Asiapac, Singapore.

Marshall, R. (1993), Storm from the East, From Genghis Khan to Khubilai Khan, University of California Press, Berkeley, Los Angeles.

May, T. (2016), Mongol Conquest Strategy in the Middle East, IN: The Mongols Middle East – Continuity and Transformation in Ilkhanaid Iran, De Nicola, B., and Melville, C. (Eds), Brill, Leiden.

Metcalfe, D. (2009), Out of Steppe, Arrow Books, London.

Mikaberidze, A. (2011), Conflict and Conquest in the Islamic World: A Historical Encyclopedia, ABC CLIO, Oxford, England.

Morgan, D. (1982), Who ran the Mongol Empire? JRAS 1982, 124–36.

Morgan, D. (1986), 'The "Great Yasa of Chingiz Khan" and Mongol Law in die Ilkhanate', *Bulletin of the School of Oriental and African Studies,* xlix, pp. 169–70.

Morgan, D. (1989), The Mongols and the eastern Mediterranean, Mediterranean Historical Review, 4:1, pp. 198–211.

Morgan, D. (2009), The Decline and Fall of the Mongol Empire, JRAS, Series 3, 19, 4, pp. 427–437.

Pacey, A. (1990), Technology in World Civilisation: A Thousand Year History, The MIT Press, Cambridge, Massachusetts.

Prawdin, M. (2017), The Mongol Empire: Its Rise and legacy, Routledge.

Richard, J. (1969), The Mongols and the Franks, Journal of Asian History, Vol. 3, No. 1, pp. 45–57.

Rogers, J. D. (2015), Empire dynamics and Inner Asia. In J. Bemmann & M. Schmauder (Eds.), Complexity of interaction along the Eurasian steppe zone in the first millennium (pp. 73–88). Bonn: Bonn University Press.

Rossabi, M. (2009), Khubilal Khan: His Life and Times, University of California Press, Berkeley.
Saunders, J. (1971), The History of the Mongol Conquests, University of Pennsylvania Press, Philadelphia.
Schurmann, H. (1956), Mongolian Tributary Practices of the Thirteenth Century, Harvard Journal of Asiatic Studies, Vol. 19, No. 3/4, pp. 304–389.
Sicker, M. (2000), The Islamic World in Ascending: From the Arab Conquests to the Siege of Vienna, Praeger Publishers, Westport.
Smith, J. (1970), Mongol and Nomadic Taxation, Harvard Journal of Asiatic Studies, Vol. 30, pp. 46–85.
Smith, J. (1975), Mongol Manpower and Persian Population, Journal of the Economic and Social History of the Orient, Vol. 18, No. 3, pp. 271–299.
Smith, J. (2000), Dietary Decadence and Dynastic Decline in the Mongol Empire, Journal of Asian History, Vol. 34, No. 1.
Skrynnikova, T. (2011), Mongolian Nomadic Society of the Empire period, IN: Alternatives to Social Evolution, Kradin, N., Korotayev, A., Bondarenko, D. (Eds), LAP Lambert Academic Publishing.
Steinhardt, N. (1983), The Plan of Khubilai Khan's Imperial City, Artibus Asiae, Vol. 44, No. 2/3, pp. 137–158.
Sweeney, J. (1982), Thomas of Spalato and the Mongols: A Thirteenth-Century Dalmatian View of Mongol Customs, Florilegium, Vol. 4, pp. 156–183.
Vernadsky, G. (1938), The Scope and Contents of Chingis Khan's Yasa, Harvard Journal of Asiatic Studies, Vol. 3, No. 3/4, pp. 337–360.
Waley-Cohen, J. (2004), The New Qing History, Radical History Review, Issue 88, pp. 193–206.
Weatherford, J. (2004), Genghis Khan and the Making of the Modern World, Crown Publishers, New York.
Wolfe, A. (2014), Marco Polo: Factotum, Auditor, Language and Political Culture in the Mongol World Empire, Literature Compass, 11/7, pp. 409–422.
Wylie, T. (1977), The First Mongol Conquest of Tibet Reinterpreted, Harvard Journal of Asiatic Studies, Vol. 37, No. 1, pp. 103–133.
Yagodin, V., and Betts, A. (2006), Ancient Khorezm, UNESCO.

10

The British Empire: 1603 AD to 1997 AD

The Europe of the 1st millennium AD could be segmented into three geographical regions.[1] The segment around the Mediterranean was cultured and civilised with professional armies, literature, banking, taxation, architecture, and philosophy.[2] This region and some areas around the Rhine and the Danube were under Roman rule. On the other hand, the peoples who were settled in northern and central Europe were relatively late comers to history and were known as the Germani.[3] The Germani represented many tribes, some much smaller than others, with distinct cultural and linguistic similarities. The Saxons were a small but significant part of this tribal grouping.[4] The Rhine Valley was the dividing line between the Gauls to the west and the Germani to the north.[5] Although the Germani occupied an area from the Netherlands to western Russia.[6]

[1] Heather, P. (2010), Empires and Barbarians: The Fall of Rome and the Birth of Europe, Oxford University Press, Oxford.

[2] Ibid.

[3] Todd, M. (2004), The Early Germans, Blackwell Publishing, Oxford, UK.

[4] Ibid.

[5] Ibid.

[6] Ibid.

S. Ramesh, *The Rise of Empires*, https://doi.org/10.1007/978-3-030-01608-1_10

The Romans recognised that the Gauls were attracted to the civilised ways of life, but the Germani were not, and remained more primitive and savage than any of the other barbarians the Romans had ever known.[7] However, compared to the peoples of eastern Europe, those living near the Rhine and the Danube were more into small-scale farming, although they did not have a civilisation to speak of.[8] Moreover, they had no known written language, so any knowledge of these Germanic peoples has been accumulated through their contact with the Romans.[9]

Rome itself fell to the Visigoths, a Germanic tribe, in 410 AD.[10] The Roman Empire began withdrawing from Britain in 388 AD, with the withdrawal being completed by 410 AD.[11] The remaining Romano-British were unable to defend themselves against attacks by the Scots and the Picts. For the defence of their frontiers, the Romano-British hired Germanic Saxons, who eventually conquered Celtic England between 446 AD and 454 AD.[12] The Saxon tribes, who came to Britain from what is now present-day Germany, after the departure of the Romans, possessed the liberal democratic values of all the ancient Germanic tribes.[13] Private property rights existed from the earliest Anglo-Saxon times, whereby individuals were entitled to 'ownership' of land and goods that they had acquired either through their own industry or through conquest.[14] However, while ownership had to be kept within the family, if the owner had no living descendants, the title of ownership could be passed to another for money in the act of sale.[15] Furthermore, although

[7] Ibid.

[8] Heather, P. (2010), Empires and Barbarians: The Fall of Rome and the Birth of Europe, Oxford University Press, Oxford.

[9] Olsen, B. (2007), Sacred Places Europe: 108 Destinations, Consortium of Collective Consciousness.

[10] Moorhead, S., and Stuttard, D. (2010), AD410 The Year That Shook Rome, The British Museum Press, London.

[11] Kubesh, K., McNeil, N., and Bellotto, K. (2006), The Romans in Britain, In the Hands of a Child, Coloma, MI.

[12] Stenton, F. (1971), Anglo-Saxon England, Third Edition, Oxford University Press, Oxford.

[13] Dolan, N, (2009), Emerson's Liberalism, The University of Wisconsin Press, Madison, Wisconsin.

[14] Thrupp, J. (1862), The Anglo-Saxon Home: A History of the Domestic Institutions and Customs of England, from the 5th to the 11th century, Longman, Green, Longman & Roberts.

[15] Ibid.

the Anglo-Saxon kingdoms of England tended to form a centralised state, there was an acknowledgement of individual freedom as long as the individual fulfilled his obligations to the state in the context of military service and the payment of tax to the King.[16] This Anglo-Saxon sense of individual freedom survived the Norman conquest of 1066 and the imposition of a feudal society. However, the Norman conquest of England concentrated power in the hands of King William and the Norman knights, who formed a new aristocracy. Thus, the Normans introduced a more robust form of feudalism into England, control of land being cemented with the building of castles in which Norman knights with royal titles to nearby lands lived.[17] Nevertheless, despite the stronger central control required to maintain a tight grip over a conquered people, the framework of the Anglo-Saxon state and values continued to survive.[18] The continual desire for the Anglo-Saxon sense of freedom and the need to restrict the power of the King evidenced itself when King John was forced to accept the terms of the nobility, in the context of the Magna Carta in 1215.[19] The Magna Carta can be seen as the source of the liberties which are expected in contemporary western societies.[20] Moreover, perhaps the roots of the need for the freedom that sprouted in Great Britain's American colonies in the latter part of the eighteenth century AD are to be found in the socioeconomic culture of the Anglo-Saxon kingdoms which emerged in the fifth century AD in England. The Saxons of England were themselves from the Germanic stock of tribes of northern and central Europe. However, there is no concrete evidence to suggest that the 'freedom' recognised by the Anglo-Saxons of England was culturally transposed from their ethnic homelands.[21]

[16] Faith, R. (1997), The English Peasantry and the Growth of Lordship, Leicester University Press, New York.

[17] Thomas, H. (2006), The Norman Conquest: England and William the Conqueror, Rowman & Littlefield Publishers Inc, New York.

[18] Stenton, F. (1971), Anglo-Saxon England, Third Edition, Oxford University Press, Oxford.

[19] Linebaugh, P. (2008), The Magna Carta Manifesto, Liberties and Commons for all, University of California Press, Los Angeles.

[20] King, F. (2015), A Magna Carta for All Humanity: Homing in on Human Rights, Routledge, New York.

[21] Harris, S. (2003), Race and Ethnicity in Anglo-Saxon Literature, Routledge, New York.

From total domination of England by the fifth century AD, the Anglo-Saxon kingdoms developed over time. The strongest of these kingdoms was Wessex. Athelstan, the grandson of King Alfred the Great of Wessex, was the first king of a unified England, reigning from 924 AD to 939 AD.[22] Athelstan was also able to successfully defend England from invasion by the inhabitants of northern Britain, Ireland, and Scandinavia.[23] He achieved this by uniting all of the Saxon kingdoms under his rule and authority. Like his father, King Edward, Athelstan was able to expand the boundaries of his English kingdom.[24] Furthermore, Athelstan developed and implemented a centralised administrative and legal system for England.[25] His courtiers and his soldiers ensured that his subjects showed obedience towards him as well as the laws he proclaimed.[26] During the early years of Alfred the Great of Wessex, one king united the Picts and the Scots in what is now Scotland, but then was known as the Kingdom of Alba. The first king to rule over both the Picts and the Scots, as is known, was King Kenneth MacAlpin I who ruled the Kingdom of Alba from 843 AD to 858 AD.[27]

According to Henry VIII, a sovereign nation is an empire when it is not subservient to either another sovereign state or ruler.[28] Later in time, the term 'empire' became applicable to a 'merger' between two states. But the roots of the British Empire lie with the granting of a charter to the East India Company in 1600, giving it monopoly rights in trade with the kingdoms of the Indian subcontinent and China. The East India Company was a joint stock company in which investors pooled their savings, giving the Company the financial resources to put together a fleet

[22] Foot, S. (2011), Athelstan – The First King of England, Yale University Press.

[23] Ibid.

[24] Ibid.

[25] Ibid.

[26] Ibid.

[27] Armstrong, D. (2002), The Kingdom of Scotland in the Middle Ages, 400–1450, Heinemann Educational Publishers, Oxford.

[28] Firth, C. (1918), The British Empire, The Scottish Historical Review, Vol. 15, No. 59, pp. 185–189.

that would sale to the East Indies with gold bullion and other items to trade for spices, silk, and textiles. Once the fleet returned to Britain with these goods, they were sold for a tidy profit, which was distributed amongst the investors according to their initial investment share. The East India Company was Britain's greatest innovation, the world's first corporation, which would marshal financial resources beyond the reach of the British state, laying the foundations of the British Empire: the mightiest and biggest empire the world has ever known. In 1603, when James I became King of England and Scotland, an empire was created.[29] To recognise this, James I as King issued a proclamation in 1604, in which he identified himself as the King of Great Britain.[30] At the same time, at least two other names were used to refer to the 'union' between England and Scotland, the 'British Empire' and the 'Britannic Empire'. However, it was not until 1707 that the kingdoms of England and Scotland legally achieved the denomination of Great Britain, that Britain's emerging colonies in the Americas began to be included within the definition of either the 'British Empire' or the 'Britannic Empire'.[31] By 1763, the population of Great Britain had reached 8 million, in addition to which there were 20 million Indians in India under British rule, 2.5 million people in the American colonies, and 2 million people in Ireland.[32] In this case, by 1765, the use of the words 'British Empire' had become more commonplace,[33] such that by 1774, King George III began to use the term to refer to his dominions.[34] However, in the context of the British Empire, the focus of analysis will remain with the region which has left its mark on world history, the America's, and a region which has seen one of humanity's oldest civilisations—the Indus Valley civilisation—the Indian subcontinent.

[29] Ibid.

[30] Ibid.

[31] Ibid.

[32] Ibid.

[33] Grigg, J. (2008), British Colonial America: People and Perspectives, ABC-CLIO, Oxford, England.

[34] Ibid.

The Americas

The British were not the first European power to attempt to colonise North America. But, eventually it was one of the most dominant of the colonisers. The Spanish established a colony at St Augustine in Florida in 1565. They also established a Jesuit mission on the Chesapeake in the 1570s, although all of the mission was later massacred by the local Indians.[35] The British also tried to establish a colony on Roanoke Island in 1587, but the colonists had disappeared when relief British ships arrived with fresh supplies. Nevertheless, the first British colony to be successfully founded in the Americas was Jamestown in Virginia in 1607.[36] The colonists only survived by trading with the local Indians, and so, did not see it as necessary to plant crops during their first year. The colonists in Jamestown only survived the first winter because the local Indians shared their food resources with the colonists.[37] Powhatan, the local Indian chief, perhaps wanted access to English goods, and so, allowed the settlement to survive, hoping at the same time to limit its expansion.[38] The establishment of the first colony in Jamestown, was followed by the establishment of 12 more colonies in subsequent years. These colonies did not share a common identity, despite being British until 1789.[39] But the beginnings of each colony had been a royal charter. This provided each colony with a system of law based on English Common Law, as well as a framework for governance.[40] Nevertheless, a common feature of these colonies was that the colonists were predisposed to enjoying gradual economic and political liberty.[41] The primary—perhaps only—purpose of the Jamestown colony was to provide a profit to the English investors who had financially supported its founding.[42] However,

[35] Grigg, J. (2008), British Colonial America: People and Perspectives, ABC-CLIO, Oxford, England.

[36] Ibid.

[37] Ibid.

[38] Ibid.

[39] Ibid.

[40] Stoebuck, W. (1968), Reception of English Common Law in the American Colonies, 10 Wm. & Mary L. Rev. 393.

[41] Ibid.

[42] Ibid.

the Chesapeake region possessed neither valuable jewels or gold. But it did have plenty of fertile land, which would support the farming of tobacco.[43] This was a valuable crop commodity which could be profitably sold back in Britain. Despite having plenty of fertile land, the New World did not have enough people to work the land and farm the tobacco. The settlers themselves were indentured labourers.[44] Thus, the only way in which the land could be productively and profitably farmed with tobacco was to kidnap people from Africa; and use them as slave labour. By 1700, there were 6000 of such people in Virginia.[45] Furthermore, some indentured English labourers could acquire land after their term of indenture had been completed. These new landowners were themselves doing well by the 1680s.[46] This signal of prosperity acted as an incentive for more English settlers to leave the shores of England for the shores of the New World. Thousands of Puritan families settled in New England between 1630 and 1642.[47] However, it is impossible to tell the accuracy of such migration figures because the requisite statistics are lacking.[48] Nevertheless, the Massachusetts Bay colony established itself much faster and was more stable than the earlier Chesapeake English colony, though still facing the same harsh conditions.[49] The relative success and stability of the New England colony may have been due to the more temperate climate that prevented the establishment of lethal diseases, which had afflicted other colonies before it.[50] Furthermore, unlike the Chesapeake, the land was not flat contiguous areas, but in small parcels surrounded by forests. This made New England unsuitable for tobacco plantations.[51] The Puritans were the radicals of the age; because they were not able to reform the

---

[43] Ibid

[44] Ibid

[45] Ibid.

[46] Ibid.

[47] Grigg, J. (2008), British Colonial America: People and Perspectives, ABC-CLIO, Oxford, England.

[48] Fogelman, A. (1992), Migrations to the Thirteen British North American Colonies, 1700–1775: New Estimates, Journal of Interdisciplinary History, XXII:4, pp. 691–709.

[49] Grigg, J. (2008), British Colonial America: People and Perspectives, ABC-CLIO, Oxford, England.

[50] Ibid.

[51] Ibid.

Anglican faith and the Church of England, they were both a threat to the state as well as a challenge to the authority of the King.[52] The settlements of the Puritans was made up of homogenous family groups.[53] This, allied with a temperate climate, allowed for their population to increase at a much faster rate than the population of other non-Puritan settlements. On the other hand, the Chesapeake settlement was made up, in comparison, of a larger number of indentured men.[54] However, the Puritans were free and from the relatively prosperous strata of English society.[55] These differences between the Puritan settlements and the non-Puritan settlements may explain the stability of the former in contrast to the latter. The expansion of the English Puritan population in the New World threatened the existence of the settlements of other European powers such as the Dutch. The Dutch, who had been experiencing a 'Golden Age' in the seventeenth century sensed the profits which could be made in the New World. This followed the return of Henry Hudson from the New World—who had been hired by the Dutch East India Company to find a sea passage to India and Asia.[56] In this case, the Dutch West India Company did not follow the English model of exploitation of the land. Instead, the Dutch, through the Company, wanted to engage in fur trading with the local Indians. And so, in order to facilitate this, they established a trading post in the Hudson River Valley at Albany in 1620.[57] Later, a second colony was established on Manhattan Island, eventually coming to be known as New Netherland, it became home to a diverse range of migrants ranging from the Jewish, Portuguese Brazilians, Quakers, and Lutherans.[58] New Netherland was taken over by the English in 1664, with the English Puritan population of the New World increasing.[59]

[52] Ibid.

[53] Ibid.

[54] Ibid.

[55] Ibid.

[56] Ibid.

[57] Ibid.

[58] Ibid.

[59] Grigg, J. (2008), British Colonial America: People and Perspectives, ABC-CLIO, Oxford, England.

Although North America saw an increase in the number of English settlers after 1607, the French also had an eye on the North American lands. However, the English colonists fought off the French with British help; and they were finally defeated in 1763. By then, the seeds of the conflict between the English colonists and the British crown had been laid, but it did not begin in earnest until 1765.[60] On the other hand, the beginnings of Canada as a British territory had its origins in the granting of control of the seas and fisheries off the coast of New Foundland to the New Foundland Company in 1610 by James I of England.[61] It was on the basis of commercial interest in fishing and, later, the fur trade, that British colonies would develop in Canada for the next 150 years.[62] It was only later that the British state would provide the formal framework in the context of governance, infrastructure, and diplomacy, which would provide the means for state formation in Canada.[63] To English merchants, the fisheries of New Foundland offered large profits at home and, for the British Economy, it generated employment as well as provided training.[64] Placing great value on the fisheries of New Foundland, Charles I then declared the 'Western Charter' in 1634, which legitimised the rights of all English subjects to be able to fish in the seas off New Foundland.[65] However, due to increasing competition from other European nations for access to the seas off New Foundland, it was recognised that the seas would be more secure if the land of New Foundland was settled by the English. Competition was mainly with the French, with the Spanish having conceded their claims to North America—with the signing of the Treaty of Madrid in 1667 in exchange for the British curtailing the activities of pirates.[66] On

[60] Ibid.

[61] Reid, J., and Mancke, E. (2008), From Global Processes to Continental Strategies: The Emergence of British North America to 1783, In Buckner, P. (Eds), Canada and the British Empire, Oxford University Press, Oxford.

[62] Ibid.

[63] Reid, J., and Mancke, E. (2008), From Global Processes to Continental Strategies: The Emergence of British North America to 1783, In Buckner, P. (Eds), Canada and the British Empire, Oxford University Press, Oxford.

[64] Ibid.

[65] Ibid.

[66] Ibid.

the other hand the French were intent on consolidating their North American territories into 'New France'.[67] The restoration of the English monarchy in 1660 gave added momentum to Britain's overseas commercial pursuits, with the formation of the Royal African Company and the Carolina Proprietors in 1663.[68] Furthermore, an English fleet defeated the Dutch and took their colony of the New Netherlands, renaming it New York, in 1664.[69] The Hudson's Bay Company was formed in 1670 with rights to Rupert's Land—around Hudson's Bay—in the context of the fur trade.[70] In order to encourage English settlers to go, the New Foundland Act was passed in 1699 in order to give grants of land to people with land claims dating to before 1685.[71] Colonies in New Foundland became more viable as agricultural surpluses from British settlements in Virginia and New England could be traded for fish.[72] The Treaty of Utrecht in 1713 further consolidated English claims in Canada, when the French agreed to English sovereignty over New Foundland and Rupert's Land.[73] The English also gained control of Acadia from the French, renaming it Nova Scotia.[74] However, this was not an easy task, and took decades of fighting before Acadia was brought under British control in 1758.[75] The war with the French between 1689 and 1713, served as the mechanism for England's ascendancy in the Americas.[76] The Treaty of Utrecht was not a single treaty, but a number of sub-treaties between the warring European states. However, it was a critical part of Britain's strategy of preventing the

[67] Ibid.

[68] Ibid.

[69] Ibid.

[70] Ibid.

[71] Ibid.

[72] Ibid.

[73] Ibid.

[74] Ibid.

[75] Grenier, J. (2014), The Far Reaches of Empire: War in Nova Scotia, 1710–1760, University of Oklahoma Press, Norman.

[76] Reid, J., and Mancke, E. (2008), From Global Processes to Continental Strategies: The Emergence of British North America to 1783, In Buckner, P. (Eds), Canada and the British Empire, Oxford University Press, Oxford.

ascendancy of a European superpower, Spain.[77] Its main European power neutered, Britain's only competitor to the formation of a global hegemony was France. The French had formed Louisiana in 1699, as well as establishing Louisbourg following the Treaty of Utrecht in 1713. Following the 1707 Act of Union between England and Scotland with a single parliament in Westminster, the power of the British state became more concentrated and powerful than it had ever been. It had access to more resources to project stronger military power abroad. Tensions had been simmering between France and Britain over control of North American trade. These tensions came to a head between these powers at a global level during the Seven Years War between 1756 and 1763.[78] The Seven Years War can be seen as the first global war, spanning not only the Americas, but also India, the Philippines, West Africa, and Europe.[79] The ending of the war in Britain's favour laid the foundations of the British Empire, with France relinquishing its control of territorial possessions in West Africa, India, and the Americas. However, the War wasn't just between the British and the French, but also included their European allies. For example, on Britain's side was Prussia, Portugal, Hanover, and the smaller German states.[80] On the other hand, France's allies included the Russian Empire, Spain, Sweden, and the Holy Roman Empire under the leadership of Austria.[81] The reason for the war was that, after the Spanish War of Succession, Silesia had been ceded from Austria to Prussia, and Austria wanted its territory back.[82] Britain, on the other hand, was worried that France would seize its ally, Hanover, in which case Britain would be forced to exchange its overseas territories for it.[83] It was for this reason that Britain and Russia negotiated a treaty in 1755, the basis of which was for Britain

[77] Elliott, J. (2014), The Road to Utrecht: War and Peace, In Dadson, T. (Ed), Britain, Spain and the Treaty of Utrecht, 1713–2013, Modern Humanities Research Association and Routledge, New York.

[78] Marston, D. (2012), The Seven Years War, Routledge, New York.

[79] Ibid.

[80] Ibid.

[81] Ibid.

[82] Calloway, C. (2006), The Scratch of a Pen:1763 and the Transformation of North America, Oxford University Press, Oxford.

[83] Ibid.

to pay Russia a sum of £100,000 per annum, for troops and ships, plus an additional payment of £400,000 if Russian troops were stationed in Hanover.[84] The French surrendered the city of Montreal in Canada to the British in September 1760.[85] However, the Seven Years War was not formally brought to an end until France, Britain, and Spain signed the Treaty of Paris in February 1763.[86] Britain's strategy had been a simple one: to—under William Pitt as Secretary of the Southern Department—reduce France to a continental power by taking away its overseas territories.[87] This strategy took on two connotations. The first was to use the British Army and its colonial militia to subjugate French forces abroad, and to use continental allies to engage them in Europe.[88] The gradual impact of this strategy was to reduce France's military capability, so that it could not contain the juggernaut of British overseas expansion. Following the defeat of General Braddock's army in Ohio in 1755, on the way to capture Fort Duquesne from the French, the British military did not encounter a major defeat such as that again. The years 1757 and 1759 represented a turning point for Great Britain. In hindsight, these years and, in particular, 1759, could be seen as the year after which Great Britain did not have to look back on its destiny to forge the greatest empire the world has ever known, the British Empire. In 1757, Robert Clive defeated the French army, and its Indian allies, in India, at the Battle of Plassey. This allowed Britain to dominate Bengal and then to gradually gain control over the whole of the Indian subcontinent.[89] In 1759, the British Army and Navy enjoyed a string of victories over the French, not only—though mainly—in the Americas, the Caribbean, as well as Africa.[90] In America, the French were gradually defeated following their abandonment of Fort Duquesne in

[84] Ibid.

[85] Ibid.

[86] Ibid.

[87] Ibid.

[88] Ibid.

[89] Haykin, M. (2012), Just before Judson, in Duesing, J. (Ed), Adoniram Judson: A Bicentennial Appreciation of the Pioneer American Missionary, B&H Publishing Group, Nashville, Tennessee.

[90] Calloway, C. (2006), The Scratch of a Pen:1763 and the Transformation of North America, Oxford University Press, Oxford.

Ohio, following Britain's Treaty of Easton with the Ohio Indians.[91] In July 1759, Fort Niagara fell to the British, followed by General Wolfe's conquest of Quebec in September of that year. The British Navy decimated the French Atlantic fleet in November 1759, and no French reinforcements or supplies ever reached the French Army or colonies in Canada.[92] 'New France' in the Americas effectively ended with the fall of Montreal to the British in 1760. In 1759, British forces also took the sugar-rich island of Guadeloupe from the French in the Caribbean, as well as the slave station of Goree in Africa.[93] France's inability to counter the British was perhaps due to the inefficiency and corruptness of the French state, which led to a waste of resources, and not only to a lack of them.[94] In a desperate attempt to prevent total British victories in North America, the Spanish allied themselves with the French in 1762.[95] However, this move cost the Spanish dearly, with Great Britain conquering Havana in Cuba as well as Manila in the Philippines, where Indian sepoys were also used in the fight.[96] The Spanish were also removed from Florida, but remained entrenched in Louisiana and New Orleans.[97] The removal of the French from North America was seen by the English colonists, whose birth rate and energy were high, as a signal that they could acquire more lands by crossing the Appalachian mountains.[98] However, Great Britain clamped down on this by requiring that no private person should buy land from the native Indians; and if any such occasion did occur, then the purchaser would be the British state.[99] This land proclamation was the first in a number of steps taken by the British state and Crown that would antagonise and cause growing resentment amongst the colonists who saw the land as

---

[91] Ibid.

[92] Ibid.

[93] Ibid.

[94] Ibid.

[95] Ibid.

[96] Ibid.

[97] Kennedy, D., Cohen, L., Piehl, M. (2017), The Duel for North America, Cengage Learning, Boston, USA.

[98] Ibid.

[99] Ibid.

their birth right.[100] The terms of the Treaty of Paris of 1763 was that all French territories in North America would be ceded to Great Britain, except for the islands of St Pierre and Miquelon in the Gulf of St Lawrence.[101] However, Great Britain returned Guadeloupe, Martinique, and St Lucia to France in exchange for French territories in Canada.[102] Guadeloupe was economically important to the French. At the time, the value of Guadeloupe's sugar exports amounted to £6,000,000.[103] Perhaps, it was the loss of France's empire in the Americas, especially with the fall of Quebec and Montreal, which laid the seeds for Great Britain to lose its colonies in America.[104] This may have been due to a few factors: firstly, the defeat of the French was not just due to the British military, but also due to the contribution of the colonial militia who, as a result, increased confidence in their own abilities.[105] Secondly, Great Britain subscribed to the theory of Mercantilism, in which stronger states should subdue weaker ones that were then forced to trade under less than equal terms with the strong state. Mercantilism caused currency shortages in the British American colonies because the colonists imported more goods from Great Britain than they exported to it.[106] The British state and Crown subjugated the freedoms of the colonists to the extent of protecting Mercantilism. For example, the production of finished iron products in the American colonies was prohibited in 1750.[107] Lastly, Great Britain had always treated its colonial subjects as inferiors in the mother country-colony partnership.[108] These factors contributed to the growing resentment and alien-

[100] Ibid.

[101] Vaughan, F. (2003), The Canadian Federalist Experiment: From Defiant Monarchy to Reluctant Republic, McGill-Queens University Press, Montreal and Kingston.

[102] Ibid.

[103] Calloway, C. (2006), The Scratch of a Pen: 1763 and the Transformation of North America, Oxford University Press, Oxford.

[104] Kennedy, D., Cohen, L., Piehl, M. (2017), The Duel for North America, Cengage Learning, Boston, USA.

[105] Ibid.

[106] Ibid.

[107] Morgan, E. (2013), The Birth of the Republic 1763–1789, Fourth Edition, The University of Chicago Press, Chicago and London.

[108] Kennedy, D., Cohen, L., Piehl, M. (2017), The Duel for North America, Cengage Learning, Boston, USA.

ation of the American colonists from Great Britain, and would eventually lead to the formation of the American Republic. The ideal and principle of freedom had become encapsulated from the very birth of the American colonies because of the gradual—and later widespread—property ownership by the settlers, in contrast to those they had left behind in the Old World.[109] Widespread property ownership by the settlers gave them not only economic, but also political, freedom.[110] The colonies themselves formed through private enterprise, albeit under licence from the British Crown.[111] The expectations of freedom by the settlers in Great Britain's American colonies—and the lack of it under the British Crown—put the colonies and the British state and Crown on a collision course, which would see the American colonists, under the Continental Congress, sign the Declaration of Independence from Great Britain in 1776.[112] The American colonies had become reliant on the import of British consumer goods, which had been increasingly being produced by small firms in England following an upturn in the English economy in 1690.[113] The boom in English production was a result of England being at the centre of a global quasi-common market centred on its overseas colonies.[114] In this quasi-common market, Britain imposed unfavourable tariffs on imported goods from its colonies, but allowed its goods to enter colonial markets at a reduced tariff.[115] This first boom in English manufacturing production was then very different from that which occurred at the time of the English Industrial Revolution from 1790 to 1870. This second boom in English manufacturing production was entirely due to declining relative costs and prices.[116] During the first boom in English production, more goods were

[109] Morgan, E. (2013), The Birth of the Republic 1763–1789, Fourth Edition, The University of Chicago Press, Chicago and London.

[110] Ibid.

[111] Ibid.

[112] Williams, B. (2002), The Declaration of American Independence, 4th July 1776, Cherrytree Books, Slough, UK.

[113] Breen, T. (1988), 'Baubles of Britain': The American and Consumer Revolutions of the Eighteenth Century, Past & Present, No. 119, pp. 73–104.

[114] Price, J. (1989), What Did Merchants Do? Reflections on British Overseas Trade, 1660–1790, The Journal of Economic History, Vol. XLIX, No. 2.

[115] Ibid.

[116] Ibid.

supplied by Great Britain to its thirteen American colonies than vice versa. In this case, Great Britain always ran a trade deficit with the colonies.[117] Goods were supplied by English merchants on credit. This may have been due to a shortage of paper and coin money in the United States.[118] But the English were content to raise as much revenue as possible from the American colonists through the imposition of duties on the sale of British imports.[119] For the colonists, the Tea Act of 1773 proved to be the final straw. Unlike other goods from England, tea was consumed in most American colonial homes.[120] The economic situation of the colonies at the time was also extremely precarious. A credit crisis, which had started in London, made it difficult for the colonists to acquire goods on credit; colonists' debts were accumulating and tobacco prices were weak.[121] The latter largely affected the southern colonies more, because it was in the southern colonies that tobacco cultivation was centred.[122] The colonial economic environment of the time and the imposition of import duties by the British made the colonial economic environment even more calamitous. The English duty on imported tea, therefore served to be the spark which ignited the path of the thirteen colonies to a revolutionary war of independence against the English.[123] The English politician and writer, John Wilkes was an ardent supporter of the American revolutionaries, having recognised the rights and the grievances of the American colonies against the King of Great Britain and his ministers.[124]

[117] Gwyn, J. (1980) British government spending and the North American colonies 1740–1775, The Journal of Imperial and Commonwealth History, 8:2, 74–84.

[118] Weiss, R. W. (1970), The Issue of Paper Money in the American Colonies, 1720–1774, J. Econ. Hist, 30 (4), pp. 770–84.

[119] Breen, T. (1988), 'Baubles of Britain': The American and Consumer Revolutions of the Eighteenth Century, Past & Present, No. 119, pp. 73–104.

[120] Ibid.

[121] Sheridan, R. (1960), The British Credit Crisis of 1772 and the American Colonies, The Journal of Economic History, Vol. XX, No. 2.

[122] Shepherd, J., and Williamson, S. (1972), The Coastal Trade of the British North American Colonies, 1768–1772, The Journal of Economic History, Vol. XXXIII, No. 4.

[123] Breen, T. (1988), 'Baubles of Britain': The American and Consumer Revolutions of the Eighteenth Century, Past & Present, No. 119, pp. 73–104.

[124] Maier, P. 91963), John Wilkes and American Disillusionment with Britain, The William and Mary Quarterly, Vol. 20, No. 3, pp. 373–395.

Great Britain fought to keep its colonies and the revolutionary war did not end until 1783, when Great Britain recognised the sovereignty of an independent United States of America, under the second Treaty of Paris, which the thirteen former colonies had joined to become one nation.[125] However, the citizens of the new country did not agree on a constitution until 1787, and the first President of the new Republic, George Washington, was not elected until 1789.[126] Hopes were renewed of a forced reconciliation between Great Britain and its former American colonies following the start of the war of 1812.[127] But the war was settled by the Treaty of Ghent; and a new diplomatic order prevailed between Great Britain and its former American colonies. Furthermore, in the years that followed the birth of the American Republic, the collapse of Mercantilism led to a surge in American exports to the colonies of the European powers in South America and the Caribbean.[128]

The Indian Subcontinent

The British East India Company began its life as a commercial and imperial vehicle for Britain's ambitions in Asia, or the East Indies, as it was known at the time.[129] Although, it had gone through several organisational changes during its 258 years of existence, it will be referred to as the 'Company'. Never, and never since, has a company had the extent and the depth of economic, political, and cultural influence which the British East India Company possessed.[130] Whereas some economists have asserted that the emergence of the multi-divisional firm in 1920s America

---

[125] Williams, B. (2002), The Declaration of American Independence, 4th July 1776, Cherrytree Books, Slough, UK.

[126] Ibid.

[127] Adelman, J., and Aron, S. (1999), From Borderlands to Borders: Empires, Nation-States, and the Peoples In between North American History, The American Historical Review, Vol. 104, No. 3, pp. 814–841.

[128] Coatsworth, J. (1967), American Trade with European Colonies in the Caribbean and South America, 1790–1812, The William and Mary Quarterly, Vol. 24, No. 2, pp. 243–266.

[129] Bowen, H. (2000), 400 Years of the East India Company, History Today, 50, 7.

[130] Ibid.

was a major organisational innovation, they had failed to grasp the idea that the British East India Company was an earlier historical example of the multi-divisional organisation of a business enterprise.[131] Moreover, in order to function, the Company had to evolve into a multi-divisional enterprise because, as a monopoly, holding sole British trading rights in the East Indies, it was truly a big enterprise.[132] A big enterprise functions more effectively if different aspects of its business are separated by function. This is very similar to Adam Smith's concept of the division of labour. The Company was granted a royal charter by Queen Elizabeth I in 1600 to conduct the business of trade with China, the Indian subcontinent, and East Asia in general.[133] However, an act of Parliament in 1698 created a New East India Company, with a provision that the New Company should give the government a £2 million loan.[134] The 'New Company' was given a monopoly on the trade of Asia beyond the Cape of Bona Esperanza up to the Straits of Magellan.[135] This effectively stopped the slave trade out of Madagascar, although it did resume briefly for a five-year period from 1716.[136] The old Company became known as the 'London Company', and it was allowed to buy into the £2 m loan, such that it gained access to a sixth of the trade of the 'New Company'.[137] The London Company and the New East India Company were effectively merged in 1709 in order to form the United Company of Merchants of England trading to the East Indies.[138] Although, still known superficially as the British East India Company, the 'merged' organisation was required to make an interest free loan to the government of £1.2 m.[139]

---

[131] Anderson, G., McCormick, R., and Tollison, R. (1983), The Economic Organisation of the English East India Company, Journal of Economic Behaviour and Organisation, 4, pp. 221–238.

[132] Ibid.

[133] Bowen, H. (2000), 400 Years of the East India Company, History Today, 50, 7.

[134] Platt, V. (1969), The East India Company and the Madagascar Slave Trade, The William and Mary Quarterly, Vol. 26, No. 4, pp. 548–577.

[135] Ibid.

[136] Ibid.

[137] Ibid.

[138] Ibid.

[139] Platt, V. (1969), The East India Company and the Madagascar Slave Trade, The William and Mary Quarterly, Vol. 26, No. 4, pp. 548–577.

The Company established its first factory in Surat, on the Indian subcontinent, in 1608.[140] Surat was a small city in Gujarat, western India, which had been incorporated into the Mughal Empire in the sixteenth century.[141] The Company was authorised to trade in Mughal India by the Emperor Jahangir in 1617.[142] However, at the same time, the Company also had factories in Japan, Siam, and the Malay peninsula, although these factories were not financially viable and shut down by 1623.[143]

The reason for the emergence of the Company, the first joint stock company in the world, was that as British trade expanded overseas, more and more financial resources were required to build fortified trading posts in foreign lands as well as to protect the trading ships that would plough the seas.[144] Before the emergence of the British East India Company, the regulated company was prominent in British commerce. The regulated company was merely an association of merchants within a specific sector that facilitated and controlled trade within that sector.[145] It was constituted solely of traders and partnerships which did not have the requisite resources to be able to deal with Britain's expanding trade frontier.[146] However, while, in contrast to the regulated company, the British East India Company was able to garner greater financial resources through the pooling of investments, in the mid-seventeenth century, the Company came under severe commercial pressure from the Dutch that it had to shift its area of focus to South Asia.[147] Moreover, following the Battle of Plassey in 1757, years of rivalry between the British East India Company

[140] Ansari, U. (2017), The Mega Year Book 2018: Current Affairs and General Knowledge, Third Edition, Disha Experts, New Delhi.

[141] Maloni, R. (2008), Europeans in Seventeenth Century Gujarat: Presence and Response, Social Scientist, Vol. 36, No. 3/4, pp. 64–99.

[142] Ansari, U. (2017), The Mega Year Book 2018: Current Affairs and General Knowledge, Third Edition, Disha Experts, New Delhi.

[143] Bassett, D. (1960), The Trade of the English East India Company in the Far East, 1624–84, Journal of the Royal Asiatic Society, Vols. 1–4, pp. 32.

[144] Baladouni, V. (1983), Accounting in the Early Years of the East India Company, The Accounting Historians Journal, Vol. 10, No. 2, pp. 63–80.

[145] Ibid.

[146] Ibid.

[147] Bowen, H. (2000), 400 Years of the East India Company, History Today, 50, 7.

and the French East India Company were ended. However, it was the Treaty of Paris in 1763 which gave Great Britain the upper hand in India. The terms of the Treaty were such that France was given back its pre-war trading posts on the Coromandel coast, but Great Britain gained exclusivity over Bengal.[148] Robert Clive's victory allowed the British East India Company to dominate Bengal and, in subsequent years, it facilitated Britain's conquest of the whole of the Indian subcontinent. This started with more and more encroachments in Bengal. For example, in 1760, the Nawab of Bengal, Mir Qasim, signed over to the Company control of Chittagong, Burdwan, and Midnapore for its support of his installation as Nawab over his father-in-law, Mir Jafar.[149] However, the demands and actions of the Company were not in the best interests of Mir Qasim and Bengal. In this case, Mir Qasim joined forces with the Nawab Wazir of Oudh and the Mughal Emperor Shah Alum to wage war against the Company.[150] But the combined forces of all three were defeated by the troops of the Company, led by Hector Munro, at the Battle of Buxar in 1764.[151] The conflict was ended with the Treaty of Allahabad in 1765, under which the Mughal Emperor ceded Bengal. Bihar and Orissa to the Company.[152] The Battle of Buxar was pivotal in changing the power structures in India, reorienting power away from Indian rulers towards the Company and the British. In this context, it was the Battle of Buxar that laid the foundations of the British Empire rather than the Battle of Plassey, by destroying the power of the Mughal Empire and its local power structures.[153] Robert Clive went on to become Governor of Bengal from 1764 to 1767.[154] During the time of his Governorship, Clive not

[148] Sutherland, L. (1947), The East India Company and the Peace of Paris, The English Historical Review, Vol. 62, No. 243, pp. 179–190

[149] Serajuddin, A. (1971), The Origin of the Rajas of the Chittagong Hill Tracts and Their Relations with the Mughals and the East India Company in the Eighteenth Century, Journal of the Pakistan Historical Society, 19, 1.

[150] Chaurasia, R. (2002), History of Modern India 1707 AD To Up to 2000 AD, Atlantic Publishers & Distributors, New Delhi.

[151] Ibid.

[152] Ibid.

[153] Ibid.

[154] Lenman, B., and Lawson, P. (1983), Robert Clive, The 'Black Jagir', And British Politics, The Historical Journal, 26, 4, pp. 801–829.

only limited the extent of the powers of the Company to Bengal, Orissa, and Bihar, but he also carried out administrative changes which allowed the Company to assume the power to tax in Bengal.[155] These diplomatic and administrative changes brought about by Clive during his Governorship transformed the Company from a mere trading organisation to a sovereign power in the Indian context.[156] Nevertheless, British belief in the governance of Bengal was that Indian institutions were adapted to Indian needs so, there was no need to supplant such institutions with British or European ones.[157]

Since then, the Company's main areas of focus on the Indian subcontinent included Bombay, Madras, and Calcutta.[158] The need to defend its areas of business from local rulers meant that the Company employed a private army, which grew to such an extent that it was comparable to the regular British Army in terms of capability.[159] However, a shortage of manpower in India necessitated the supplementation of the Company's troops with British regulars in 1754.[160] Moreover, the Company's troop contingent was supplemented with regular units of the British Army and Navy, especially around the times of the Austrian War of Succession and the Seven Years War.[161] In 1749, there had only been 3000 troops guarding the Company's settlements in India but, by 1778, the number of the Company's troops in India had swollen to 67,000 men.[162] This force not only included British troops, but also native Indian sepoys who had been trained in British military methods.[163] Furthermore, between 1802 and

[155] Ibid.

[156] Bowen, H. (1987), Lord Clive and Speculation in East India Company Stock, 1766, The Historical Journal, 30, 4, pp. 905–920.

[157] Marshall, P. (2011), The British Presence in India in the 18th Century: Retrieved 5th May 2018 from BBC: History: http://www.bbc.co.uk/history/british/empire_seapower/east_india_01.shtml

[158] Bowen, H. (2000), 400 Years of the East India Company, History Today, 50, 7.

[159] Ibid.

[160] Gilbert, A. (1975), Recruitment and Reform in the East India Company Army, 1760–1800, Journal of British Studies 15, pp. 89–111.

[161] Sutherland, L. (1947), The East India Company in Eighteenth-Century Politics, The Economic History Review, Vol. 17, No. 1, pp. 15–26.

[162] Bryant, G. (1978), Officers of the East India Company's army in the days of Clive and Hastings, The Journal of Imperial and Commonwealth History, 6:3, 203–227.

[163] Ibid.

1858, 90,000 trained British recruits travelled from the Company's barracks on the Isle of Wight to India, to join the Company's army as private soldiers.[164] The Company was then able to use military force in order to protect trade, and was, therefore, known as a fighting company.[165] In this context, the activities and the contributions of men like Robert Clive, Charles Cornwallis, and Warren Hastings from a trading house to a sovereign power in its own right, having mastery over the lives of millions of Indians by the mid-eighteenth century.[166] The fortunes of the Company and the British economy were closely tied. In this case, the Company can be seen as an institution which laid the framework for business and, eventually, sovereign transactions between Great Britain and the peoples of the Indian subcontinent.[167] The returning ships of the Company, with their valuable cargoes of tea, spices, and textiles would dock in London, where the Company had not only its headquarters, but also vast warehouses for storing cargoes before the goods could be distributed. The Company's growth required more and more administrative staff to organise its business in India as well as trained military personnel to protect such business. This was especially true in the context of its commercial rivalry with the French in the Indian subcontinent.[168] So, the Company established a college to train its future administrators and military personnel at Haileybury, and a Military Seminary in Croydon, respectively.[169] Moreover, the Company maintained administrative offices throughout Great Britain.[170] The Company raised its financial resources from the sale of its shares and bonds in the emerging British stock market.[171] The Company increased its profitability by changing its business

[164] Mokyr, J. (1996), Height and Health In the United Kingdom 1815–1860: Evidence from the East India Company Army, Explorations in Economic History, 33, 7, pp. 141–168.

[165] Bowen, H. (2000), 400 Years of the East India Company, History Today, 50, 7.

[166] Ibid.

[167] Stern, P. (2009), History and Historiography of the English East India Company: Past, Present and Future, History Compass, 7/4.

[168] Gilbert, A. (1975), Recruitment and Reform in the East India Company Army, 1760–1800, Journal of British Studies 15, pp. 89–111.

[169] Bowen, H. (2000), 400 Years of the East India Company, History Today, 50, 7.

[170] Ibid.

[171] Ibid.

model from one of simply exchanging British goods and bullion for Indian commodities, to a business model based on the acquisition of land in north-east India and the collection of taxes.[172] This was because the elasticity of demand for British goods in India was very low because consumer tastes were rigid.[173] The result was that the Company was having to exchange precious bullion rather than British goods for Indian goods.[174] However, by the latter half of the eighteenth century, the Company was also beginning to monopolise the textiles industry in Bengal. This was achieved by instituting a legal framework which prohibited local weavers from working for private merchants, if the weaver accepted cash advances for orders from the Company.[175] Furthermore, the weavers were driven into debt when the Company returned textile products deemed to be poor quality.[176] Local weavers could not produce replacements due to their ongoing delivery schedule.[177] As a result, more and more local weavers became indebted to the company and either had to take flight to another region to work as weavers or stay where they were and move into other sectors of the local economy, such as cultivation.[178] The Company relied on the legal framework and on enforceable contracts in order to go about its business.[179] In the case of employees, contracts were formulated such that employees were incentivised to work on behalf of the company, but at the same time morally obligated to do their duty. This kind of behaviour was instituted by allowing employees not only to trade on behalf of the Company directors, but also for themselves, with the added threat of dismissal.[180]

---

[172] Ibid.

[173] Chaudhuri, K. (1963), The East India Company and the Export of Treasure in the Early 17th century, The Economic History Review, New Series, Vol. 16, No. 1, pp. 23–38.

[174] Ibid.

[175] Hossain, H. (1979), The Alienation of Weavers: Impact of the Conflict between the Revenue and Commercial Interests of the East India Company, 1750–1800, The Indian Economic and Social History Review, Vol. XVI, No. 3.

[176] Ibid.

[177] Ibid.

[178] Ibid.

[179] Hejeebu, S. (2005), Contract Enforcement in the English East India Company, The Journal of Economic History, Vol. 65, No. 2.

[180] Ibid.

By collecting taxes in India and using it to pay for British goods, the British East India Company not only changed Britain's terms of trade with the subcontinent, but also stopped the use of British bullion to pay for British goods.[181] This helped the British economy to accumulate wealth. The right to collect taxes had been given to the Company by the Mughal Emperor Shah Alam II in 1765, in its stronghold of Bengal as well as in the neighbouring states of Orissa and Bihar.[182] The British Treasury also received a share of the revenues accumulated by the Company in Bengal in return for allowing the Company to administer Bengal and to collect tax revenues in its own right.[183] Thus, the British state financially benefited without having to expend any resources whatsoever.[184] The arrangement was a boon to the British state, which was expending money to subdue and control its American colonies, which was leading to an ever-increasing national debt.[185] However, the Company was unable to generate the sort of revenues which would help its investors to thrive and pay the British state due to its increasing complex role in India, which required the use and expenditure of more and more resources. By 1772, the Company was on the verge of bankruptcy and unable to deliver either the required returns to its investors or its contribution to the coffers of the British state.[186] As a result, in an attempt to increase revenues for the British state in other ways, the exclusive rights of the Company were increasingly eroded as time went by. The importance of the Company to the British state and economy was underlined by the need to make it more accountable to Parliament, which would only serve to make it more efficient.[187] This was achieved by the passing of the Regulation Act of 1773 and the India Act of

[181] Robins, N. (2002), Loot: in search of the East India Company, the world's first transnational corporation, Environment & Urbanisation, Vol. 14, No. 1.

[182] Bowen, H. (2000), 400 Years of the East India Company, History Today, 50, 7.

[183] Ibid.

[184] Ibid.

[185] Ibid.

[186] Ibid.

[187] Ibid.

1784.[188] For example, its monopoly over trade in India was ended in 1813 and its exclusive right to trade with China was ended in 1833.[189] The ending of the Company's monopoly on trade with India, underlined Great Britain's emergent commitment to free trade.[190] The passing of the East India Company Charter Act of 1813, allowed merchants outside of Great Britain and the British East India Company to take part in trade between Britain, India, and Southeast Asia.[191] The Industrial Revolution that began in England in the late eighteenth century resulted in an ever-increasing supply of goods, which required more selling opportunities in new markets.[192] Finally, following the Indian Mutiny of 1857, the British state took over administration of India; and the British Raj was born and the Company died. However, throughout colonial times, the British in India were shaped by India.[193] This began in Calcutta, Great Britain's administrative capital of its appropriated land in Bengal from 1772.[194] It was here that British and Indian society first mixed. Bilateral societal knowledge was exchanged through informal (relationships, trade) connections rather than through formal (schools) connections.[195] Through such informal and, later, formal connections, Britain would establish an elite Indian class that would help it to administer an ever-expanding geographical reach on the subcontinent. However, after the 1830s more formal establishments were set up to train surveyors, administrators, accountants, and medical assistants.[196] Over time, Great Britain would eventually unite the disparate kingdoms of India into one country.

---

[188] Ibid.

[189] Ibid.

[190] Webster, A. (1990), The Political Economy of Trade Liberalisation: The East India Company Charter Act of 1813, The Economic History Review, New Series, Vol. 43, No. 3, pp. 404–419.

[191] Ibid.

[192] Mitchener, K., and Weidenmier, M. (2008), Trade and Empire, The Economic Journal, 118.

[193] Marshall, P. (1997), British Society in India under the East India Company, Modern Asian Studies, Vol. 31, No. 1, pp. 89–108.

[194] Marshall, P. (2000), The White Town of Calcutta under the Rule of the East India Company, Modern Asian Studies, Vol. 34, No. 2, pp. 307–331.

[195] Ibid.

[196] Pernau, M. (2012), India in the Victorian Age, Victorian India? IN Victorian World, Hewitt, M. (Ed), Routledge, New York.

Britain's Other Colonies

At its height, the British Empire included Dominions (Canada, Australia, New Zealand, the Cook Islands, Union of South Africa, Rhodesia, and British India), colonies in Asia (including Malaysia, Singapore, Borneo, Ceylon, and Hong Kong), colonies in Africa (including Nigeria, Gambia, Kenya, Uganda, and Egypt), and colonies in the Caribbean (Trinidad, Bahamas, Barbados, and British Guiana).[197] The opening of trading posts in Penang in 1786 and Singapore in 1819 heralded Great Britain's arrival on the Malaysian peninsula.[198] The institutions which the British brought with them to Malaysia, property rights, the rule of law, and individual freedom ushered in opportunities for entrepreneurship as well as the chance to make a fortune. This acted as a magnet for attracting a range of peoples from Southeast China, India, West Asia, Indonesia, and Europe, creating a melting pot of nationalities.[199]

China and Britain had been on par in terms of technological ability until the start of the eighteenth century. However, towards the latter half of the eighteenth century, this tilted in the favour of Great Britain, with the start of the Industrial Revolution in England. This allowed England to become more productive than any other country in the world through technological and organisational innovation.[200] It therefore needed more resources, which it could acquire by dominating other peoples through Mercantilism, using its far superior military forces. Until 1833, the British East India Company could trade with China through the port of Canton.[201] However, the British found trade with China expensive because they had to pay for Chinese tea and ceramics in silver.[202] Britain

[197] Mitchener, K., and Weidenmier, M. (2008), Trade and Empire, The Economic Journal, 118.

[198] Gin, O. (2009), Politics Divided: Malaysia-Singapore Relations, In Shiraishi, T. 9Ed), Across the Causeway: A Multi-Dimensional Study of Malaysia-Singapore Relations, Institute of Southeast Asian Studies, Singapore.

[199] Ibid.

[200] Tsang, S. (2007), A Modern History of Hong Kong, I.B.Tauris & Co Ltd., London.

[201] Gelber, H. (2016), Battle for Beijing, 1858–1860, Franco-British Conflict in China, Palgrave Macmillan.

[202] Ibid.

gained confidence from its naval victory at Trafalgar in 1815, and it was fast losing patience with China.[203] However, it grew opium in India and then shipped it to China, where it was sold to the local population. This drained China of vast amounts of silver.[204] It fought the First Opium War with China between 1840 and 1842, ending the war with a treaty with China on highly unfavourable terms to that country. These terms included the granting of Most Favoured Nation trading status to Britain, but at the same time being reciprocated to other European powers.[205] Furthermore, Britain required that its consuls be stationed at trading ports in China, and that China should pay reparations to Britain.[206] However, while Great Britain and the other European powers were able to weaken China, they were not able to acquire China as a colony, as they had done in other parts of the world. The reason that the European powers were not able to subjugate China was that, by the time of their arrival in the nineteenth century, China had become a vast, unified country, ruled by the Qing Dynasty since 1644.[207] Britain reverted control of its last remaining territory in Asia, Hong Kong, back to the People's Republic of China in 1997.

References

Adelman, J., and Aron, S. (1999), From Borderlands to Borders: Empires, Nation-States, and the Peoples In between North American History, The American Historical Review, Vol. 104, No. 3, pp. 814–841.

Anderson, G., McCormick, R., and Tollison, R. (1983), The Economic Organisation of the English East India Company, Journal of Economic Behaviour and Organisation, 4, pp. 221–238.

Ansari, U. (2017), The Mega Year Book 2018: Current Affairs and General Knowledge, Third Edition, Disha Experts, New Delhi.

[203] Ibid.

[204] Ibid.

[205] Ibid.

[206] Ibid.

[207] Smith, R. (2015), The Qing Dynasty and Traditional Chinese Culture, Rowman & Littlefield, London.

Armstrong, D. (2002), The Kingdom of Scotland in the Middle Ages, 400–1450, Heinemann Educational Publishers, Oxford.
Baladouni, V. (1983), Accounting in the Early Years of the East India Company, The Accounting Historians Journal, Vol. 10, No. 2, pp. 63–80.
Bassett, D. (1960), The Trade of the English East India Company in the Far East, 1624–84, Journal of the Royal Asiatic Society, Vols. 1–4, pp. 32.
Bowen, H. (1987), Lord Clive and Speculation in East India Company Stock, 1766, The Historical Journal, 30, 4, pp. 905–920.
Bowen, H. (2000), 400 Years of the East India Company, History Today, 50, 7.
Breen, T. (1988), 'Baubles of Britain': The American and Consumer Revolutions of the Eighteenth Century, Past & Present, No. 119, pp. 73–104.
Bryant, G. (1978), Officers of the East India Company's army in the days of Clive and Hastings, The Journal of Imperial and Commonwealth History, 6:3, 203–227.
Calloway, C. (2006), The Scratch of a Pen: 1763 and the Transformation of North America, Oxford University Press, Oxford.
Chaudhuri, K. (1963), The East India Company and the Export of Treasure in the Early 17th century, The Economic History Review, New Series, Vol. 16, No. 1, pp. 23–38.
Chaurasia, R. (2002), History of Modern India 1707AD To Up to 2000AD, Atlantic Publishers & Distributors, New Delhi.
Coatsworth, J. (1967), American Trade with European Colonies in the Caribbean and South America, 1790–1812, The William and Mary Quarterly, Vol. 24, No. 2, pp. 243–266.
Dolan, N. (2009), Emerson's Liberalism, The University of Wisconsin Press, Madison, Wisconsin.
Elliott, J. (2014), The Road to Utrecht: War and Peace, In Dadson, T. (Ed), Britain, Spain and the Treaty of Utrecht, 1713–2013, Modern Humanities Research Association and Routledge, New York.
Faith, R. (1997), The English Peasantry and the Growth of Lordship, Leicester University Press, New York.
Firth, C. (1918), The British Empire, The Scottish Historical Review, Vol. 15, No. 59, pp. 185–189.
Fogelman, A. (1992), Migrations to the Thirteen British North American Colonies, 1700–1775: New Estimates, Journal of Interdisciplinary History, XXII:4, pp. 691–709.
Foot, S. (2011), Athelstan – The First King of England, Yale University Press.
Gelber, H. (2016), Battle for Beijing, 1858–1860, Franco-British Conflict in China, Palgrave Macmillan.

Gilbert, A. (1975), Recruitment and Reform in the East India Company Army, 1760–1800, Journal of British Studies 15, pp. 89–111.

Gin, O. (2009), Politics Divided: Malaysia-Singapore Relations, In Shiraishi, T. (Ed), Across the Causeway: A Multi-Dimensional Study of Malaysia-Singapore Relations, Institute of Southeast Asian Studies, Singapore.

Grenier, J. (2014), The Far Reaches of Empire: War in Nova Scotia, 1710–1760, University of Oklahoma Press, Norman.

Grigg, J. (2008), British Colonial America: People and Perspectives, ABC-CLIO, Oxford, England.

Gwyn, J. (1980) British government spending and the North American colonies 1740–1775, The Journal of Imperial and Commonwealth History, 8:2, 74–84.

Harris, S. (2003), Race and Ethnicity in Anglo-Saxon Literature, Routledge, New York.

Haykin, M. (2012), Just before Judson, in Duesing, J. (Ed), Adoniram Judson: A Bicentennial Appreciation of the Pioneer American Missionary, B&H Publishing Group, Nashville, Tennessee.

Heather, P. (2010), Empires and Barbarians: The Fall of Rome and the Birth of Europe, Oxford University Press, Oxford.

Hejeebu, S. (2005), Contract Enforcement in the English East India Company, The Journal of Economic History, Vol. 65, No. 2.

Hossain, H. (1979), The Alienation of Weavers: Impact of the Conflict between the Revenue and Commercial Interests of the East India Company, 1750–1800, The Indian Economic and Social History Review, Vol. XVI, No. 3.

Kennedy, D., Cohen, L., and Piehl, M. (2017), The Duel for North America, Cengage Learning, Boston, USA.

King, F. (2015), A Magna Carta For All Humanity: Homing in on Human Rights, Routledge, New York.

Kubesh, K., McNeil, N., and Bellotto, K. (2006), The Romans in Britain, In the Hands of a Child, Coloma, MI.

Lenman, B., and Lawson, P. (1983), Robert Clive, The 'Black Jagir', And British Politics, The Historical Journal, 26, 4, pp. 801–829.

Linebaugh, P. (2008), The Magna Carta Manifesto, Liberties and Commons for all, University of California Press, Los Angeles.

Maier, P. (1963), John Wilkes and American Disillusionment with Britain, The William and Mary Quarterly, Vol. 20, No. 3, pp. 373–395.

Maloni, R. (2008), Europeans in Seventeenth Century Gujarat: Presence and Response, Social Scientist, Vol. 36, No. 3/4, pp. 64–99.

Marshall, P. (1997), British Society in India under the East India Company, Modern Asian Studies, Vol. 31, No. 1, pp. 89–108.

Marshall, P. (2000), The White Town of Calcutta under the Rule of the East India Company, Modern Asian Studies, Vol 34, No. 2, pp. 307–331.

Marshall, P. (2011), The British Presence in India in the 18th Century: Retrieved 5th May 2018 from BBC: History: http://www.bbc.co.uk/history/british/empire_seapower/east_india_01.shtml

Marston, D. (2012), The Seven Years War, Routledge, New York.

Mitchener, K., and Weidenmier, M. (2008), Trade and Empire, The Economic Journal, 118.

Mokyr, J. (1996), Height and Health In the United Kingdom 1815–1860: Evidence from the East India Company Army, Explorations in Economic History, 33, 7, pp. 141–168.

Moorhead, S., and Stuttard, D. (2010), AD410 The Year That Shook Rome, The British Museum Press, London.

Morgan, E. (2013), The Birth of the Republic 1763–1789, Fourth Edition, The University of Chicago Press, Chicago and London.

Olsen, B. (2007), Sacred Places Europe: 108 Destinations, Consortium of Collective Consciousness.

Pernau, M. (2012), India in the Victorian Age, Victorian India? IN Victorian World, Hewitt, M. (Ed), Routledge, New York.

Platt, V. (1969), The East India Company and the Madagascar Slave Trade, The William and Mary Quarterly, Vol. 26, No. 4, pp. 548–577.

Price, J. (1989), What Did Merchants Do? Reflections on British Overseas Trade, 1660–1790, The Journal of Economic History, Vol. XLIX, No. 2.

Reid, J., and Mancke, E. (2008), From Global Processes to Continental Strategies: The Emergence of British North America to 1783, In Buckner, P. (Eds), Canada and the British Empire, Oxford University Press, Oxford.

Robins, N. (2002), Loot: in search of the East India Company, the world's first transnational corporation, Environment & Urbanisation, Vol. 14, No. 1.

Serajuddin, A. (1971), The Origin of the Rajas of the Chittagong Hill Tracts and Their Relations with the Mughals and the East India Company in the Eighteenth Century, Journal of the Pakistan Historical Society, 19, 1.

Shepherd, J., and Williamson, S. (1972), The Coastal Trade of the British North American Colonies, 1768–1772, The Journal of Economic History, Vol. XXXIII, No. 4

Smith, R. (2015), The Qing Dynasty and Traditional Chinese Culture, Rowman & Littlefield, London.

Stenton, F. (1971), Anglo-Saxon England, Third Edition, Oxford University Press, Oxford.

Stern, P. (2009), History and Historiography of the English East India Company: Past, Present and Future, History Compass, 7/4.

Stoebuck, W. (1968), Reception of English Common Law in the American Colonies, 10 Wm. & Mary L. Rev. 393.

Sutherland, L. (1947a), The East India Company and the Peace of Paris, The English Historical Review, Vol. 62, No. 243, pp. 179–190.

Sutherland, L. (1947b), The East India Company in Eighteenth-Century Politics, The Economic History Review, Vol. 17, No. 1, pp. 15–26

Thomas, H. (2006), The Norman Conquest: England and William the Conqueror, Rowman & Littlefield Publishers Inc, New York.

Thrupp, J. (1862), The Anglo-Saxon Home: A History of the Domestic Institutions and Customs of England, from the 5th to the 11th century, Longman, Green, Longman & Roberts.

Todd, M. (2004), The Early Germans, Blackwell Publishing, Oxford, UK.

Tsang, S. (2007), A Modern History of Hong Kong, I.B.Tauris & Co Ltd, London.

Vaughan, F. (2003), The Canadian Federalist Experiment: From Defiant Monarchy to Reluctant Republic, McGill-Queens University Press, Montreal and Kingston.

Webster, A. (1990), The Political Economy of Trade Liberalisation: The East India Company Charter Act of 1813, The Economic History Review, New Series, Vol. 43, No. 3, pp. 404–419.

Weiss, R. W. (1970), The Issue of Paper Money in the American Colonies, 1720–1774, J. Econ. Hist, 30 (4), pp. 770–84.

Williams, B. (2002), The Declaration of American Independence, 4th July 1776, Cherrytree Books, Slough, UK.

11

Conclusion

The purpose of this book has been to evaluate the role of innovation in economics, the factors it depends upon, and how innovation emerges over time. The latter is achieved based on a historical analysis of several civilisations over a 10,000-year periods, but starting from the emergence of humanity from its origins in the savannahs, jungles, and forests of Africa. At this time, invention and innovation would have been driven by the motivation to survive, and much of it would have been driven by instinctive intelligence. For example, that it would be safer to sleep high up in the trees at night than on the ground, and that being in large groups would be safer than remaining solitary. However, the evolutionary development of cognitive intelligence only occurred in Homo Sapiens. This cognitive intelligence represents the ability to think, invent, and innovate based on environmental information received via sight, hearing, feeling, and smell. For example, Homo Erectus would have seen fire, heard it, felt it, and smelt it. At the same time, Homo Erectus may have come across deer or antelope 'cooked' by wildfire and tasted it. It would then have put its limited cognitive capacity to use by harnessing fire, by creating the conditions required to start it and to cook the animals it had caught or the kills it had taken from other animals. The cognitive capacity of Homo Erectus would have been limited compared to its successor,

S. Ramesh, *The Rise of Empires*, https://doi.org/10.1007/978-3-030-01608-1_11

Homo Sapiens, principally because the brains of Homo Erectus young stopped growing after birth.[1] Nevertheless, the brains of Homo Sapiens young continued to develop years after birth, giving Homo Sapiens far more cognitive capacity than that achieved by Homo Erectus. The latter unknowingly aided this evolutionary development by eating cooked meat. Eating cooked meat facilitated the evolutionary transition of the large body and small brain of Homo Erectus, to the smaller body and larger brain, packed with many more neurons, of Homo Sapiens.[2] Nevertheless, the role of neurons in intelligence and creativity is being actively challenged.[3] In this case, new evidence suggests that the glia, the matter between neurons, perform two functions in brain physiology.[4] Firstly, the glia can replace themselves, as well as neurons acting as stem cells. Secondly, the glia may act as stores of information. It has also been found that higher intelligence requires a higher ratio of glia to neurons. Thinking triggers the firing of neurons in the brain. This process will connect the different regional glia, allowing for different information stores to be accessed simultaneously, and for thoughts to form. Experiences and observations will be connected through a thought process, allowing insights to form, with creativity and innovation following. Therefore, intentional actions, such as learning, may be best at directing the pace of technological change in some phases of innovation than in others.[5]

Cooked meat provided far more calories for Homo Erectus than would a day of foraging, so large bodies were no longer required, in evolutionary terms.[6] The greater cognitive capacity of early Homo Sapiens made the species more inquisitive of its surroundings as well as being more aware

[1] Coqueugniot, H., Hublin, J., Veillon, F., Houet, F., and Jacob, T. (2004), Early Brain Growth in Homo Erectus and Implications for Cognitive Ability, Nature, 431, 299–302.

[2] Fonseca-Azeuedo, K., and Herculano-Houzel, S. (2012), Metabolic Constraint Imposes Trade-off Between Body Size and the Number of Brain Neurons in Human Evolution, PNAS, 109 (45) 18571–18576; doi: https://doi.org/10.1073/pnas.1206390109

[3] Ibid.

[4] Koob, A. (2009), The Root of Thought: Unlocking Glia – The Brain Cell That Will Help Us Sharpen Our Wits, Heal Injury and Treat Brain Disease, Pearson Education, New Jersey, USA.

[5] Hughes, T.P. (1983), Networks of Power: Electrification of Western Society 1880–1930, Johns Hopkins Univ. Press, Baltimore, MD.

[6] Ibid.

of its environment than any of its Homo-related ancestors. Knowledge of the use of fire for light in darkness, heating, and cooking may have diffused through communities, as well as knowledge of primitive weapons. Early Homo Sapiens' increased cognitive capacity would have enabled it to adapt to any environment on earth by being inventive and innovative, coming up with solutions to novel problems. However, the evolutionary development of human society and economy over time suggests that the development of human cognitive capacity is a dynamic one that is still evolving. The development of human cognitive capacity continued as Homo Sapiens set forth from the African continent to colonise the rest of the earth. Homo Sapiens continued to experience a hunter-gatherer lifestyle until 9000 BC, when the first permanent Neolithic settlements emerged in ancient Sumer. Innovations such as the spindle whorl and the potter's wheel emerged during the Hassuna culture of 6000 BC to 4000 BC and the Ubaid period from 4000 BC to 3500 BC, respectively. It was at around 4000 BC that metallurgy developed such that it enabled the production of metal tools to shape and manipulate wood more effectively. Nevertheless, more sophisticated innovations, such as the two-wheel horse-drawn chariot with spoke wheels, did not appear in the region until the rule of the Kassites from 1505 BC to 1157 BC. The Kassites were the first rulers/civilisation to recognise a political organisation based on the notion of a nation state rather than upon a city state. The Assyrian Empire, 2500 BC to 609 BC, on the other hand, was the first civilisation to establish a permanent library with 20,000 cuneiform-inscribed tablets at Nineveh. Therefore, the Assyrian Empire was not just solely based on military dominance, but also through scholastic activity.

The earliest copper objects that have been unearthed in the world have been found in Anatolia from around 9000 years ago. However, the unearthed evidence of Yangshao bronze artefacts, from the Yangshao culture of China, were dated to 3000 years later than the copper artefacts found in Anatolia. So, innovation and technological development was occurring at a faster rate in the near east than it was in ancient China. Moreover, the innovation of permanent settlement did not spread to China until 4000 BC. Nevertheless, in the transition from a hunter-gather existence to a more permanently settled lifestyle, intermediate sites of worship and less permanently settled habitation such as the one at

Gobekli Tepe may have developed. Such semi-habitation of sites for worship may have provided the experiences and the learning to facilitate the innovation towards the permanent settlement of sites near rivers and the switch to crop rearing and animal husbandry from the previous hunter-gatherer existence. This occurred in the plains of ancient Mesopotamia and the Levant and, as a result, the first civilisations came into existence. However, the transition from a hunter-gatherer existence to a settled urban life in cities may have involved a semi-settled existence in Neolithic times. In this case, while people were settled in specific sites and were engaging in crop cultivation and animal husbandry, there may have been a division of labour in which some of the population were still engaged in hunting and gathering. This would have enabled the group to benefit from a diverse diet, which would have contributed to enhancing the development of the brain, not only during the foetal stage, but beyond, into childhood, adolescence, and adulthood. As a result, the increased number of neuronal connections linking each glia space would have increased cognitive capacity of Homo Sapiens over time. The species would then have been better able to innovate in order to adapt to a changing environment, and therefore, maximising survival. In this case, because the development of Homo Sapiens' cognitive capacity is a dynamic process which changes over time, so is the process of innovation. However, in order to develop, innovation is further facilitated by the accumulation of knowledge over time. The development of a written form of analysis would facilitate this. Archaeological evidence suggests that the first form of a very simplified writing system was invented in ancient Egypt at around 3250 BC. Egypt was part of the migratory route for Homo Sapiens out of Africa, but no archaeological evidence of Homo Sapiens has been found there between 11,000 BC and 8000 BC. However, the late Neolithic period between 5100 BC and 4700 BC was a period, as supported by archaeological evidence, of increased signs of human habitation especially in the western desert region. Trade facilitated the diffusion of knowledge of the early form of writing from ancient Egypt to the settlement of Uruk, located in its time to the east of the channel of the current flow of the Euphrates River. It was in this region that the Sumerian civilisation, the first in the humanity of history, emerged at around 3500 BC. The Sumerian civilisation inherited a knowledge of writing from Uruk, however, this too was centuries before the pictorial script of Uruk developed into Sumerian Cuneiform—a more sophisticated form of earlier writing that would have allowed for the transmission of more information. Therefore, the Sumerians were the peoples of the first civilisation in humanity to have a written language, which facilitated the development of written laws. Thus, a permanent record would exist through which human behaviour could be regulated and human activity more efficiently organised. In this case, the Sumerians also favoured individual achievement. The development of a more elaborate form of writing in Sumer also enabled the development of mathematics which then was based on a structure which encompassed number words. Writing in China may have developed independently over a period of time although there is the possibility that some form of trade link may have existed between the near-eastern civilisations, the Indus Valley civilisations, and that of ancient China, through trade. While writing may have developed over time, either through knowledge diffusion or independently, it was not until the late Shang period (1200 BC to 1045 BC) that the earliest inscriptions appeared on bronze objects belonging to the nobility. Furthermore, the earliest literature in China appeared during the Zhou dynasty, from 1046 BC to 256 BC. An earlier, less sophisticated culture, the Longshan culture, which extended from 3000 BC to 1900 BC, served as the roots of the early Chinese states, the Xia (2100 BC to 1600 BC), Shang (1600 BC to 1050 BC), and the Zhou (1046 BC to 256 BC). The diffusion of knowledge in the ancient world may be deducible from the fact that the people of ancient Sumer traded with the people of the Indus Valley civilisation during the Harappan phase, which extended from 2600 BC to 2500 BC. The period encompassing the Egyptian Old Kingdom (2650 BC to 2134 BC) corresponded with the sophisticated urbanisation phase of the Indus Valley civilisation, as well as with the presence of the Akkadian Empire from 2350 BC to 2100 BC. The Indus Valley civilisation was based on a hierarchy of wealth rather than on political control. Archaeological evidence suggests that the people of the Indus Valley civilisation were able to build ships with a cabin on deck, steered with an oar and a helm. Given the fact that the Chola Dynasty, which ruled, on and off, from 350 BC to 1279 AD, had trading links with both the Roman Empire and Song dynasty China, the diffusion of knowledge through trade in the ancient world may not be so implausible. Other civilisations have also contributed towards contemporary society. It was in the city states of ancient Greece that the concept of freedom centred on the citizen emerged. This was then embodied in the politic of the Roman republic, in which the centralisation of power by the unelected senate gave way to autarkic rule and the rise of the Roman Empire. Certain aspects of the technical knowledge of the Roman Empire were at the same level of sophistication as that of the Indus Valley civilisation, which had existed millennia before. The Song dynasty represents a period in which there was a rapid transformation of Chinese society in the context of demographic changes, changes in governance, learning, and innovation in association with a commercial revolution. This has led some scholars to suggest that, in the eleventh century, Song dynasty China was at the cusp of an industrial revolution, which did not occur for another 500 years until late eighteenth-century England.[7] Therefore, The Song dynasty came close to being the first civilisation in the world to undertake industrialisation had it been not for the emergence of the civil service exams and the Mongol conquest. This diffused knowledge of Chinese innovations—gunpowder, printing, paper, the compass, and a civil service—to the west, which innovated upon the accumulated knowledge. It was the military innovation and the discipline of the Mongols which enabled their prowess to defeat more innovative and established civilisations (Khorezm, China, Persia, Iran) as well as less advanced civilisations (Europe). England then innovated by developing the forebear of the modern-day corporation to establish itself as the British Empire, the last vestiges of which did not die out until the late twentieth century, although a tiny proportion remains today.

Contemporary innovation theory places an emphasis on the sources of innovation either from a demand pull or a technology push perspective. Demand-pull theory asserts that innovation arises because of market demand. However, in doing so the theory is neither able to explain how the innovative capacity changes over time or how innovation is related to

[7] Elvin, M. (1973), The Pattern of the Chinese Past: A Social and Economic Interpretation, Stanford University Press, Stanford.

market conditions. On the other hand, the Technology Push theory of innovation asserts that purposeful scientific research is the initiator of innovation. In other words, this theory asserts that the supply side of the market is more efficient at delivering innovation than is the demand side. However, the Technology Push theory of innovation fails to recognise the importance of economic factors to innovation, and the causal relationship between economic stability and the direction of technological change. In this case, studies have shown that there is a direct link between economic and institutional factors and the innovation process. For example, economic stability depends on political stability as well as the rule of law. For example, state ideology and support would foster creativity and innovation, as is the case with contemporary China. Nevertheless, while state support is an important driver of innovation, it also needs to be recognised that people and their experiences are essential for innovation and invention to occur. However, while economic theory may suggest that education and learning by doing may contribute to innovation and invention, it does not explicitly address the role of human cognition in this process. This lack of recognition of human cognition in the invention and innovation process is also alluded to in both the Actor Network theory and the Cultural-Historical Activity theory of innovation. The former is based on the idea that the development of innovation is based on the network of agents connected to it. On the other hand, the Cultural-Historical Activity theory suggests that the factors driving innovation may be unique due to cultural factors at a specific point in time. In the context of both theories, the role of the individual and human cognition is important for both invention as well as for innovation. In the context of human cognition, invention and innovation may occur because of observation, experience and trial and error. However, innovation is sustainable and cumulative when knowledge can diffuse between civilisations, as well as from one civilisation to another over time, resulting in the accumulation of a body of knowledge. In other words, invention in one time period is known about in another time period, leading to innovation. But this lends to the proposition that there is some form of continuity or stability in the economic and political environment of civilisation. This could either be in the context of a continuing civilisation, developing and growing over time, such as the transformation of

England into the British Empire, or the immersion of a civilisation by exogenous forces, as in the case of ancient Babylon. The transition of England into the British Empire involved institutional changes with the rising power of the merchant class and the dwindling power of the feudal landed gentry. The personification of this transition was the ascendancy of the doctrine of Mercantilism. In essence, this doctrine featured a balance of trade in favour of the mother country, with colonial outposts supplying only raw materials and primary production but undertaking no manufacturing of goods, which lay exclusively in the hands of the merchants of the mother country, the mother country at the time being England. This doctrine of Mercantilism allowed the merchants of England to become wealthy. However, there had been a shift away from the monopoly of the guilds towards more state control of business. The state granted monopoly trading rights to a corporation of investors in favour of financial assistance. As such, the power of the state became embedded in commercial activities. The rise of the corporation, such as the British East India company, was an English financial and organisational innovation that allowed a large group of investors to pool financial resources, and therefore, to finance far larger trading activities with the added incentive of greater individual profits. In a monetary economy, individual wealth accumulation facilitates individual and familial survival. Although wealth accumulation will allow for the financing of innovation, the notion of wealth accumulation may lead to Machiavellian behaviour. On the other hand, in a non-monetary economy, survival depends on the physical strength of the individual, on the strength of numbers as well as on technological, organisational, and procedural innovations which would give one army an advantage over another. In this case, the use by the Mongols of lightly clad archers on horseback, coupled with their terrifying reputation for ruthlessness in battle and upon defeated foe, gave them an advantage over opposing armies. Furthermore, when a civilisation was militarily weak through internal dissension or through natural disasters, it would become susceptible to conquest, as perhaps happened with the Indus Valley civilisation, which was well advanced for its time and was bigger than contemporary civilisations, such as that of ancient Mesopotamia and ancient Egypt. Although the intimate details of the Indus Valley civilisation have been lost to time, its language not yet having been deciphered either, whatever knowledge of it does exist has been obtained by archaeological digs and surveys. However, it is known that the merchant ships of the Indus Valley civilisation traded with the cities of ancient Mesopotamia, and trade would have allowed for the diffusion of knowledge from one civilisation to that of another.

The institutions which have formed in China over the last 40 years of its economic reforms, coupled with its technocratic democracy in which people have economic freedom but not the political freedom favoured in Europe, perfectly fits the economic, cultural, technological, and the societal environment of the early twentieth century. China needs its style of technocratic democracy, in which people are selected by their peers to represent them at a local level at the annual National Congress, because of its vast population. Human behaviour needs to be controlled so that the freedom for individuals to do bad is denigrated, but their freedom to do good is promoted. The analysis for this conclusion results from Fig. 1.1 in Chap. 1, where it was elucidated that the basis for human existence is to maximise survival, and not to maximise utility. The maximisation of survival would involve the interplay of a range of factors, resulting in negative and positive behaviours, whereas the maximisation of utility is unrealistic because it relies upon logic for a positive behavioural outcome. Throughout its history, China's governance has been based on autocratic rule, firstly by the dynasties and, in recent history, by one party. The static nature of change in the form of China's governance has not been matched by the emergence of different schools of thought, which include Confucianism, Taoism, and Legalism. These schools of thought have, at different times, become culturally embedded and have acted as the cement of autocratic rule. Legalism fulfils this role in modern-day China. While the people of China have economic freedom and economic prosperity, it is unlikely that they would want to change a system of governance that has delivered them these things in 40 years, in conjunction with further economic development in the future. The western democratic model—being based on freedom of choice because of the theoretical importance of utility maximisation from an economics perspective—ignores the bad behaviour—which results from economic freedom and political freedom—of individuals. Thus, this distinction between the freedom to do good and the freedom to do bad is not recognised in western-style traditional democracy. The latter was reincarnated from the ideals of ancient Greece and ancient Rome—at a time when the world population was under 1 billion people, whereas now it is over 7.5 billion people—in the late eighteenth century. It was at this time period that the peak of the doctrine of freedom also occurred, with its champion of economic discourse, Adam Smith, and the revolutionaries who declared the Republic of America in 1776.[8] But the stirrings of the re-emergence of western freedom lay in the signing of the Magna Carta in thirteenth-century England by King John, whose powers were sought to be curtailed by the nobles.[9] It was also in the thirteenth century that the Mongol armies decimated the then-global population, at a time when the estimated global population was only 360 million people.[10] Not only are China's governance institutions better able to efficiently manipulate the global economic, cultural, technological, and the societal environment of the early twentieth century, but it can do so more effectively than can the traditional western concept of freedom and democracy, with its roots in ancient Greece and ancient Egypt. Furthermore, China's economic reforms of the last 40 years and the incremental development of its governing institutions has facilitated innovation, entrepreneurship,[11] and the embracing of free trade. However, despite embracing free trade, China, like the England of the sixteenth century to the mid-nineteenth century, is also following the doctrine of Mercantilism, in which its purpose is to accumulate substantial foreign exchange reserves earned from its exports, but at the same time increasingly gaining access to and control of other countries' resources. The strategy to achieve this is to give countries more and more loans which they cannot repay in kind, but to the extent of giving long leases on strategic and economic assets. Therefore, just as England was building its empire from the late seventeenth century

[8] Williams, B. (2002), The Declaration of American Independence: 4th July 1776, Cherrytree Book, Berkshire, UK.

[9] Vincent, N. (2012), Magna Carta: A very Short Introduction, Oxford University Press, Oxford.

[10] Lopez, G. (2016), How the world went from 170 million people to 7.3 billion, in one map, https://www.vox.com/2016/1/30/10872878/world-population-map

[11] Ramesh, S. (2018), Entrepreneurship in China and India, Journal of the Knowledge Economy, https://doi.org/10.1007/s13132-018-0544-y

onwards, so too is the China of today following the same path to development and economic and military dominance. However, whereas in the context of western economic development the economic concept of the free market—where resources were best and efficiently allocated by the price mechanism through supply and demand—was closely linked with freedom of choice, freedom of expression, and political freedom, this did result in economic progress through innovation for a time when the population was small, colonisation possibilities made land not so scare—although at the expense of indigenous peoples—and the vote was restricted to the gentry and the land-owning class. However, after political suffrage became universal in western countries after the early twentieth century, this led to economic decline over subsequent decades. This was also at a time when stealing other people's land was no longer a viable option, although wars were fought to promote the western model of democratic governance. The result of increasing and widespread universal suffrage in western countries like Britain was greater economic inefficiency. This was exacerbated after the 1960s by a drop in the calibre of British politicians who were no longer tied to national goals, but to European ones, despite having been a global power at odds with other European powers for many centuries before. However, China's economic development has resulted in greater economic efficiency and greater innovation than has the rise of the west simply because of the separation of economic freedom from political freedom. China has produced flexible and dynamically changing institutions that have allowed it to become a rapidly innovating economy. More so, because its governance structures allow for the most efficient allocation of resources compared to the western model of governance, which was most appropriate at a time when the populations of the world and of nation states was very much smaller than they are today. When populations are large, central control produces better economic results and increased allocative and production efficiency because the divergence of economic freedom from political freedom allows for the market mechanism of the 'Intelligent Hand' of a strong central government to replace Adam Smith's 'Invisible Hand', in which economic freedom and political freedom were enshrined in the context of choice. The implication of the 'Invisible Hand' was that citizens enjoyed both economic and political freedom, but the latter dictated the nature of the former. However, while the 'Invisible Hand' produced better economic performance from the eighteenth century to the mid-twentieth century in Europe, it occurred at a time when the populations of these countries, and the world, was very much smaller than it is today. Today, China's form of governance is economically superior to that of the western model for two reasons. Firstly, simply because it reflects a better way to economically manage a very large population. And secondly, a strong central technocratic government is better able to allocate resources through the 'Intelligent Hand'. Moreover, while the species Homo Sapiens is essentially an enlightened animal with higher cognitive capability than other species on earth, it is still an animal whose only natural goal is to survive and pass on its genes in perpetuity. In doing so, it is capable of uncivilised behaviour that can only be countered by strong central government control that facilitates 'good' behaviours but is able to prevent 'bad' behaviours. This contrasts with the western model in which, although a distinction is made between the two kinds of behaviours, the freedom to do either is not prevented. Thus, China's form of governance is better suited to today's global environment, to produce better economic efficiency than is the contemporary western system of governance. This is particularly evident in countries such as the United States, Great Britain, and Italy—countries in which populism has arisen due to the economic deprivation of its peoples. The governance systems of these countries did contribute positively to their economic development in the past, but this is no longer the case. This is because the governance systems of these countries, as well as countries with similar governance systems, no longer efficiently allocate resources, as does China's current system of governance. However, China's economic, social, military, and political rise will lead to conflicts with the established order of western governance. But this is inevitable in the rise of empires, the next level of which will be heavily influenced by which nation is best able to harness the power of artificial intelligence for warfare.

References

Coqueugniot, H., Hublin, J., Veillon, F., Houet, F., and Jacob, T. (2004), Early Brain Growth in Homo Erectus and Implications for Cognitive Ability, Nature, 431, 299–302.

Elvin, M. (1973), The Pattern of the Chinese Past: A Social and Economic Interpretation, Stanford University Press, Stanford.

Fonseca-Azeuedo, K., and Herculano-Houzel, S. (2012), Metabolic Constraint Imposes Tradeoff Between Body Size and the Number of Brain Neurons in Human Evolution, PNAS, 109 (45) 18571–18576. doi: https://doi.org/10.1073/pnas.1206390109

Lopez, G. (2016), How the world went from 170 million people to 7.3 billion, in one map, https://www.vox.com/2016/1/30/10872878/world-population-map.

Ramesh, S. (2018), Entrepreneurship in China and India, Journal of the Knowledge Economy, doi: https://doi.org/10.1007/s13132-018-0544-y

Vincent, N. (2012), Magna Carta: A Very Short Introduction, Oxford University Press, Oxford.

Williams, B. (2002), The Declaration of American Independence: 4th July 1776, Cherrytree Book, Berkshire, UK.

Index[1]

[1]Note: Page numbers followed by 'n' refer to notes.

S. Ramesh, *The Rise of Empires*, https://doi.org/10.1007/978-3-030-01608-1

B

J

K

L

Z

The manufacturer's authorised representative in the EU is Springer Nature Customer Service Centre GmbH, Europaplatz 3, 69115 Heidelberg, Germany. If you have any concerns regarding our products, please contact ProductSafety@springernature.com

Printed and bound by CPI Group (UK) Ltd, Croydon, CR0 4YY
30/09/2026
02225656-0014